Astronomers' [

More information about this series at
http://www.springer.com/series/6960

Hugh M. Van Horn

Unlocking the Secrets of White Dwarf Stars

Hugh M. Van Horn
Alexandria, Virginia, USA

ISSN 1614-659X ISSN 2197-6651 (electronic)
ISBN 978-3-319-09368-0 ISBN 978-3-319-09369-7 (eBook)
DOI 10.1007/978-3-319-09369-7
Springer Cham Heidelberg New York Dordrecht London

Library of Congress Control Number: 2014951702

Springer is part of Springer Science+Business Media (www.springer.com)

The author with several of his students and "grand-students" at the Second Conference on Faint Blue Stars in 1987. *Front row, left to right*: Pierre Brassard, Austin Tomany, S. O. Kepler, Pierre Chayer, Jean Dupuis. *Second row:* Pierre Bergeron, Robert Lamontagne, Chuck Claver, J. Allen Hill, Matt Wood, Theo Koupelis. *Third row, standing*: Stephane Vennes, Steve Kawaler, Didier Saumon, Gilles Fontaine, the author, François Wesemael, Don Winget, Butler Hine. (Image courtesy of Steven D. Kawaler. Used with permission.)

This book is dedicated to all the men and women whose work has contributed to the advancement of our understanding of white dwarf stars.

Preface

Today, we know that the objects astronomers call "white dwarfs" are among the most common stars in the Milky Way Galaxy. Indeed, the majority of stars end their lives by becoming white dwarfs.

These points were by no means clear at the beginning of the twentieth century, however. Then, only three of these faint stars had been discovered. Each contained about as much mass as our Sun, but it was crammed into a volume only about the size of Earth. If a sample of this "star stuff" about the size of a sugar cube could have been brought into a terrestrial laboratory, it would have weighed as much as a small car!

What *were* these mysterious objects? The scientists of the day were completely baffled. Indeed, the quest to understand their nature and determine their relation to other stars spanned most of the past century. In the end, it necessitated contributions from perhaps a thousand scientists and required revolutionary advances in physics, observing capabilities, and technology. And, of course, a great deal of hard work and human ingenuity!

As a participant in this international research effort since the 1960s, I witnessed at first hand some of the detective work necessary to solve the puzzles posed by white dwarfs. The overall story has always seemed to me a fascinating one. It involved serendipitous discoveries, the interplay between observations and theory, the pivotal role of technological advances, interaction across scientific disciplines, and a protracted international effort. It also provides an excellent case study of the way science actually works. For these reasons, I felt motivated to tell the story as I came to know it, thinking that others might find it interesting as well.

Everyone who undertakes to write a history is faced with making decisions about what to put in and what to leave out, and different people will of course make different choices. By the time

I reached the middle of this book, I began to realize that some of the choices I was making were undoubtedly very different from those others might make, even if they had backgrounds similar to my own. Part of the reason is that my own research played a part in this story, and of course I included that. I also realized that I have tended to emphasize those aspects of the research done by others with which I was most familiar. While I have tried to be even-handed in telling this story, others will undoubtedly find that I have failed to emphasize—or perhaps even mention—aspects that they find more compelling. To any friends or colleagues who may feel that their contributions have not received the recognition they feel appropriate, I sincerely apologize. Any errors—either of omission or of commission—are entirely my responsibility.

In the end, I hope that this book has captured some of the sense of excitement I have felt in being privileged to take part in the effort to unlock the mysteries of white dwarfs. And I hope that readers will find the recounting of that effort to be an entertaining story of the detective work necessary to accomplish it.

Alexandria, VA, USA Hugh M. Van Horn

About the Author

Hugh M. Van Horn's entire career since 1965 has been dedicated to conducting research into the properties of dense matter and applying that knowledge to objects such as white dwarfs and neutron stars. For almost 30 years, he led a research team investigating these topics at the University of Rochester, and a number of his former students have gone on to distinguished careers in astrophysics themselves. He is personally acquainted with many of the senior leaders in white dwarf research across the globe.

His key contributions to our current understanding of the white dwarfs include (1) demonstrating that these cooling stars crystallize while still at temperatures where they are observable, (2) helping to develop detailed physical models for the properties of dense matter in their surface convection zones and deep interiors, and (3) helping to develop the theory of pulsating white dwarfs to the point where asteroseismological studies are now routinely being used to determine physical properties of these stars.

Dr. Van Horn has co-edited or co-authored several books including: (1) *White Dwarfs and Variable Degenerate Stars,* ed. H. M. Van Horn and V. Weidemann (University of Rochester: Rochester, NY, 1979); (2) *Strongly Coupled Plasma Physics, ed. H. M. Van Horn and S. Ichimaru* (University of Rochester Press: Rochester, NY, 1993); and (3) *2020 Vision: An Overview of New Worlds, New Horizons in Astronomy and Astrophysics,* H. M. Van Horn and M. Specian (booklet published by the National Academy of Sciences: Washington, D.C., 2011).

Acknowledgments

I am grateful to the late Ed Salpeter, my Ph.D. thesis advisor, for teaching me how to think as a scientist and how to do scientific research. I also thank him for introducing me to the fascinating properties of extremely dense matter, a subject with which I have since maintained a lifelong interest. In addition, I thank Malcolm Savedoff, with whom I worked as a postdoc, for introducing me to white dwarfs and to stellar evolution calculations. The late Carl Hansen, my close friend and collaborator, shared with me and our students the engrossing journey to understand the pulsating white dwarfs. My former students—especially Don Lamb, Gilles Fontaine, Don Winget, the late François Wesemael, and Didier Saumon—were my partners in working to improve our understanding of the properties of these dense stars and of the matter of which they are composed. So, too, have been their own students, whom I am proud to have as my "grand-students." Postdocs Brad Carroll, Gilles Chabrier, Jeff Colvin, and Shuji Ogata also have been companions in this effort, each bringing their own special expertise to bear on one or more aspects of the larger problem.

My friend and colleague, Jim Liebert, worked with me for many years on a draft of a so-far-unpublished technical book about white dwarfs. Jim has also made numerous and substantial contributions to the spectroscopy of the white dwarf stars. To him and to my University of Rochester colleagues Malcolm Savedoff, Larry Helfer, and Judy Pipher I am grateful for helping me to appreciate the vital importance of observations to the progress of astronomy and astrophysics. Thanks also to Malcolm, Larry, and my other fellow Rochester faculty members Eric Blackman, Adam Frank, and Jack Thomas for helping to make astrophysical theory such enjoyable work.

I further want to thank Jay Holberg for a number of stimulating discussions as he was preparing his own book, *Sirius: Brightest*

Diamond in the Night Sky, for publication. His example helped inspire me to write this book, and Jay additionally provided useful advice along the way.

I am indebted also to the late Karl-Heinz Böhm, John Cox, Jesse Greenstein, Bob Marshak, Ed Nather, Forrest Rogers, Evry Shatzman, and Volker Weidemann not only for their own substantial contributions in helping to unlock the secrets of the white dwarf stars but also for their collegial interest in my own work and for their help at various times.

Thanks, too, to numerous other friends and colleagues around the world for helping to educate me about many aspects of the properties of white dwarfs. I have had the pleasure of working on research problems with some of you, co-authored papers with others, was stimulated by the publications of still more, and delighted in listening to your oral presentations. Though I do not have the space here to thank each of you individually, I remain grateful to all of you for your friendship and inspiring work.

Many people helped me to locate images for this book. In addition to those who are credited in the figure captions, I want to thank Charles Alcock, Shirley Brignall, Matt Burleigh, Chris Clemens, Cheryl Cunningham, Allison Doane, Richard Green, Jay Holberg, Buell Januzzi, David Monet, David Nather, Eisha Neely, Kevin Quillen, George Rieke, Tyler Trine, Dan Watson, Jennifer West, Kurtis Williams, Don Winget, Barbara Wolff, and Matt Wood.

In addition, I am grateful to the National Science Foundation (NSF) for supporting my own research on white dwarfs over the years and for enabling me to devote time to working on a technical book on this subject with Jim Liebert. I have also found the Astrophysics Data System maintained by the Smithsonian Astrophysical Observatory, with support from the National Aeronautics and Space Administration (NASA), to be an invaluable resource for reference material.

I also want to thank Clive Horwood at Praxis Publishing for his prompt and efficient handling of my proposal for this book and for his continuing support. And I am especially grateful to my Springer editors, Maury Solomon and Nora Rawn, for their patience, advice, and help throughout the process of readying this book for publication. Without them, it would not have happened.

Last but by no means least, I am eternally grateful to my cherished wife, Sue, for her love and support over the years, for her patience and understanding during the writing of this book, and for her critical reading of several drafts. This book would never have happened without her.

Contents

Preface.. ix

Acknowledgments xiii

List of Illustrations.. xix

1. The First Clues 1

2. A Star the Size of the Earth? Absurd!.. 9

3. Great Balls of Fire.. 17

4. Relativity, Wave-Particle Duality, and the Nature of White Dwarfs 29

5. Star Power.. 41

6. Still Pretty Hot for a Fading Old Star!.. 53

7. Stalking the Wild White Dwarf 63

8. The Peculiar Spectra of White Dwarfs 75

9. Interlude: Crossing the Digital Divide 87

10. How to Make a White Dwarf.. 99

11. Diamonds in the Sky.. 109

12. The Envelope, Please! 123

13. Leaping into Space.. 137

14. Decoding the Spectra of White Dwarfs 151

15. The Secrets in the Spectra 165

16. Understanding the White Dwarf Menagerie 183

17. Music of the Spheres 195

18. The Whole Earth Telescope and Asteroseismology 211

19. Magnetic Personalities 225

20. Odd Couples 237

21. White Dwarfs and the Nature of the Milky Way Galaxy 247

22. White Dwarfs and Cosmology 259

Appendix A Some Useful Astronomical Units 271
Appendix B Powers of Ten and Logarithms 275
Appendix C Chandrasekhar's Models for Fully Degenerate White Dwarfs 277
Appendix D "WD Numbers" for White Dwarfs in this Book 279
Abbreviations and Symbols 281

Glossary 287

Bibliography 305

Index 309

List of Illustrations

Fig. 1.1 Friedrich W. Bessell 2
Fig. 1.2 Variations in the path of Sirius across the sky 4
Fig. 1.3 Alvan Clark 6

Fig. 2.1 Director Edward Pickering and the female staff of the Harvard College Observatory in 1913 10

Fig. 3.1 Ejnar Hertzsprung 20
Fig. 3.2 Henry Norris Russell 21
Fig. 3.3 Russell's 1914 version of the Hertzsprung-Russell diagram 23
Fig. 3.4 Arthur Stanley Eddington 26

Fig. 4.1 Albert Einstein in the patent office in Bern, Switzerland, about 1905 30
Fig. 4.2 Subramanyan Chandrasekhar, about 1939 36
Fig. 4.3 The theoretical mass-radius relation for fully degenerate white dwarfs, according to Chandrasekhar 38

Fig. 5.1 Hans A. Bethe in 1935 43
Fig. 5.2 The zero-age Main Sequence in the Hertzsprung-Russell diagram 47
Fig. 5.3 The early evolution of a 1.2 solar mass star with low abundances of heavy elements 49
Fig. 5.4 Edwin E. Salpeter 51

Fig. 6.1 Robert E. Marshak, *circa* 1950 54
Fig. 6.2 Temperature distribution in the white dwarf Sirius B, according to 1940 calculations by Marshak 56
Fig. 6.3 Leon Mestel in 1976 59

Fig. 7.1 Willem J. Luyten 64

Fig. 8.1 Gerard P. Kuiper 76
Fig. 8.2 George Ellery Hale 79
Fig. 8.3 The 200-in. Hale Telescope on Mt. Palomar 81
Fig. 8.4 Jesse L. Greenstein 81
Fig. 8.5 Sketches of a selection of white-dwarf spectra 83

Fig. 9.1 A modern reconstruction of the Atanasoff-Berry Computer, or "ABC," originally built at Iowa State College in the early 1940s 89
Fig. 9.2 ENIAC, the Electronic Numerical Integrator and Computer, built at the University of Pennsylvania in the mid 1940s 91

Fig. 10.1 The evolution of stars that become white dwarfs, from the Main Sequence to the white-dwarf stage 106
Fig. 10.2 The Ring Nebula in the constellation Lyra 107

Fig. 11.1 Zero-temperature Hamada-Salpeter models for white dwarfs 110
Fig. 11.2 Malcolm P. Savedoff 113
Fig. 11.3 Donald Q. Lamb, Jr. 116
Fig. 11.4 Cooling curve for a white dwarf model that includes crystallization but not phase separation 118

Fig. 12.1 Evry Schatzman in 1973 124
Fig. 12.2 Erika Böhm-Vitense 127
Fig. 12.3 Karl-Heinz Böhm in 1973 127
Fig. 12.4 Gilles Fontaine at the Université de Montréal in 1999 130

Fig. 12.5 Variation of the amount of mass in the convection zone of a 0.612 solar mass white dwarf with a pure helium surface layer as the luminosity of the star decreases 134
Fig. 12.6 The logarithm of the central temperature T_c as a function of the logarithm of the luminosity in solar units for a 0.612 M_{Sun} white dwarf with a helium envelope 135

Fig. 13.1 The *Copernicus* spacecraft 139
Fig. 13.2 Jay Holberg 141
Fig. 13.3 The *International Ultraviolet Explorer (IUE)* spacecraft 142
Fig. 13.4 The *Hubble Space Telescope (HST)* 145

Fig. 14.1 Volker Weidemann at a 1980 meeting in Erice, Italy 152
Fig. 14.2 Harry Shipman receving an award at the University of Delaware in 2008 156
Fig. 14.3 Detlev Koester 159
Fig. 14.4 François Wesemael at the Université de Montréal in 2003 160
Fig. 14.5 Pierre Bergeron 162

Fig. 15.1 James Liebert at a 2006 meeting in Leicester, UK 166
Fig. 15.2 Comparison between theoretical models and observations in the color-magnitude diagram for white dwarfs from the USNO's parallax program 170
Fig. 15.3 Test of the white-dwarf mass-radius relation 172
Fig. 15.4 Judi Provencal 173
Fig. 15.5 Virginia Trimble in 1988 175
Fig. 15.6 Gary Wegner 178
Fig. 15.7 The mass distribution of the DA white dwarfs 182

Fig. 16.1 Peter A. Strittmatter at a 2009 meeting in Columbus, Ohio 185
Fig. 16.2 Dayal T. Wickramasinghe during a meeting in Tübingen, Germany, in 2010 185

Fig. 16.3 Annie Baglin 187
Fig. 16.4 Gerard Vauclair at a meeting in Tübingen, Germany, in 2010 187
Fig. 16.5 The "Fingers of God" in the Eagle nebula 190
Fig. 16.6 The modified two-channel picture for the formation of white dwarfs with H- and He-dominated spectra, as proposed by Shipman in 1989 191

Fig. 17.1 Carl J. Hansen 199
Fig. 17.2 Don Winget (background) at the University of Texas in 1985 during a visit by Icko Iben, Jr. (foreground) 201
Fig. 17.3 Angular patterns of some non-radial pulsation modes 202

Fig. 18.1 R. Edward Nather 213
Fig. 18.2 The distribution of observatories that participated in the WET campaign in March 1989 215

Fig. 19.1 The spectrum of the magnetic white dwarf Feige 7 227
Fig. 19.2 A sketch of the magnetic field distribution for a model for the strongly magnetic white dwarf PG 1031+234 232

Fig. 20.1 Artist's concept of a dwarf nova system 242

Fig. 21.1 The Andromeda Galaxy 248
Fig. 21.2 Comparison between the faint end of the WDLF obtained in 1998 by Sandy Leggett and her colleagues with theoretical models by Matt Wood 251
Fig. 21.3 The globular cluster M4 253

Fig. 22.1 Saul Perlmutter, Adam Riess, and Brian Schmidt during a press conference on the occasion of winning the Nobel Prize in 2011 260

1. The First Clues

Friedrich Wilhelm Bessel (1784–1846; see Figure 1.1) discovered the first clue that pointed to the existence of a strange new class of stars, which we now call white dwarfs. Although he recognized that his finding was important, neither he nor any other scientist in the early nineteenth century had the remotest idea that dramatic advances in science and technology and a revolution in fundamental physics would be required before these stars could be understood.

At the beginning of the nineteenth century, astronomers knew almost nothing about the structures of the stars. They did know — by comparing their own measurements of the positions of stars in the sky against determinations made by Ptolemy in Egypt nearly 2,000 years earlier and those of the ancient Greeks centuries still further back into the past[1]—that some of the supposedly "fixed" stars actually moved across the sky. However, they did not know how far away they were, or how big they were, or what they were made of, and they did not know how much power a star actually radiates to produce the starlight we see on Earth.

During the first half of the nineteenth century, astronomers made all of their observations by eye directly at the telescope, as photography had not yet been invented. Astronomy was thus a demanding discipline, especially in northern Europe, where Bessel worked. A clear night was also a cold night, particularly during the

[1] For the historical information in the first five chapters, I have relied heavily on the books by Langer (1968), Pannekoek (1989), and Holberg (2007). Langer provides a very useful outline of world history, including notable advances in science and technology, from the earliest times to the middle of the twentieth century. Pannekoek similarly summarizes the history of astronomy from the time of the first historic civilizations to the latter part of the twentieth century. In addition, Chandrasekhar (1939) provides valuable historical information about the early developments in stellar astrophysics. Holberg provides a very readable account of the story of our knowledge of the white dwarf star Sirius B, starting from ancient Egypt — millenia before the existence of Sirius B was even suspected! — up to the discoveries made from spacecraft observations in the late twentieth and early twenty-first centuries.

H.M. Van Horn, *Unlocking the Secrets of White Dwarf Stars*,
Astronomers' Universe, DOI 10.1007/978-3-319-09369-7_1

FIGURE 1.1 Friedrich W. Bessel. Public domain image courtesy Wikimedia Commons (http://commons.wikimedia.org/wiki/File:Friedrich_Wilhelm_Bessel.jpeg)

winter months. It required a true passion for the science to be willing to spend night after night in a sometimes bitter-cold observatory, with an eye glued to a telescope eyepiece, to make astronomical observations.

Bessel had not started out to be an astronomer. Instead, he began working as a clerk in a merchant's office in Bremen, in the northern German state of Hanover. However, he was very much interested in navigation and astronomy, and as he studied books on these subjects his passion for astronomy grew. He eventually gave up his business position to become an assistant in Amtmann Schroeter's private astronomical observatory in Lillienthal, some 10 km from Bremen. Bessel proved to be adept in practical matters of physical measurement and computing, and he was naturally gifted in mathematics, later inventing a new class of special functions — now called "Bessel functions" — in solving a problem in planetary motions.

In 1810, the 26-year-old Bessel was selected to found a new astronomical observatory in Königsberg, located in Bavaria, near Bamberg in central Germany, some 500 km south of Bremen. There he initiated a careful reduction of a lengthy series of observations published more than half a century earlier by British astronomer James Bradley. Bessel was especially careful in analyzing the unavoidable residual errors in the observations, and his results, published in 1818, set a new, high standard for the reduction of astronomical data.

In 1820, Bessel installed a meridian circle at the observatory in Königsberg. This device consisted of a graduated circle fixed perpendicularly upon the horizontal axis of an instrument used to observe transits of stars across the meridian.[2] The arrangement allowed simultaneous measurements of the "right ascensions" and "declinations" of stars — essentially their longitudes and latitudes, respectively, on the celestial sphere — with smaller and more easily determined errors than were obtainable by earlier methods. High-accuracy readings were obtained by viewing sharply engraved division marks through a microscope and reading them with a micrometer. A telescope with a large achromatic lens attached in the middle of the axis provided sharp, rounded images, and a reticle — or cross-hairs—in the focal plane enabled precision measurements of stars down to the ninth apparent magnitude[3]. By moving the telescope slowly and steadily, an observer could make a star follow the horizontal wire of the reticle exactly, and by listening to the ticking of the observatory clock he or she could estimate in tenths of a second the moment the star passed the vertical wire. Bessel was a pioneer in the field of precision astronomy using this instrument and was recognized as one of the preeminent astronomers of his day.

Bessel's goal — and the main focus of the ablest astronomers during the early nineteenth century — was to obtain accurate positions of thousands of stars in an effort to build up a reference frame for the celestial coordinates of all stars. These measurements, and

[2] See the book's glossary for definitions of unfamiliar terms.

[3] Apparent magnitude is a numerical measure of the brightness of a star as seen from Earth. See the glossary for additional information.

similarly accurate determinations by other leading astronomers, had a number of important byproducts. One was the first measurement of the parallax of a star, the apparent shift in the position of a nearby star against the background of more distant stars produced by the orbital motion of Earth around the Sun during the year.

By measuring the parallax angle and knowing the distance from Earth to the Sun — termed the astronomical unit or AU[4] — it was possible for the first time to determine the distances to the nearest stars by triangulation. Bessel's 1840 measurements for the star 61 Cygni revealed a parallax of 0.348 seconds of arc (0.348″) with a mean error of 0.14", giving the distance to this star as 590,000 au or — in more convenient units for measuring stellar distances – 2.9 parsecs. In 1870, astronomer William L. Elkin (1855–1933), then at Yale University Observatory in New Haven, Connecticut, found the parallax of Sirius to be 0.38″, putting it at a distance of 543,000 au, or 2.63 parsecs. An angular measure of one second of arc is equivalent to the apparent diameter of a nickel at a distance of approximately 4 km, which illustrates just how demanding these measurements were.

By 1844, Bessel's very accurate positional measurements had shown that the bright stars Sirius and Procyon both exhibited irregularities in their motions across the sky (see Figure 1.2). Instead of moving steadily, as expected for an isolated star, each star oscillated about a straight-line path.

On August 10, 1844, Bessel wrote to Sir John Herschel in England about this discovery, and Herschel promptly published a translation in the *Monthly Notices of the Royal Astronomical Society*. "If we were to regard Sirius and Procyon as double stars," Bessel wrote,[5] "these changes in their motions would not surprise us..." However, if this were the correct explanation, why were the companion stars not seen, as they were in other binary systems? This was the first clue to the existence of an unusual new class of star.

Bessel died of cancer in Königsberg in March of 1846, only 2 years after his seminal discovery, and Christian August Friedrich Peters (1806–1880) succeeded him as director of the Königsberg

---

[4] See Appendix A in this book for more about units of measurement in astronomy.

[5] Holberg (2007), p. 57.

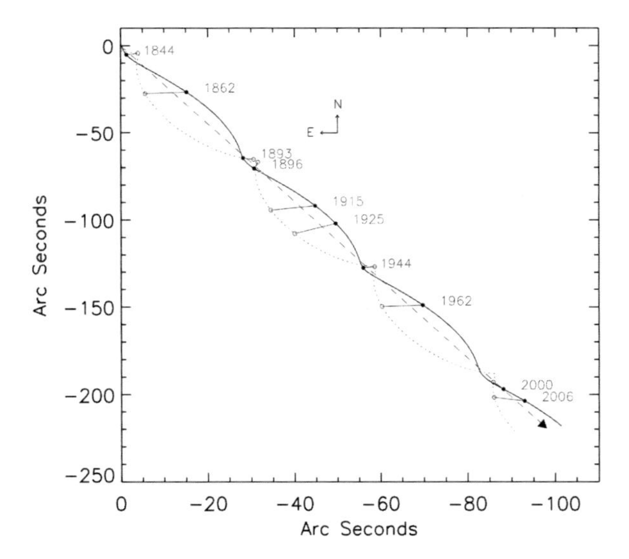

FIGURE 1.2 Variations in the path of Sirius across the sky, as charted by Jay Holberg from historical data. The *solid curve* marked by the *filled circles* represents the relative motion of Sirius (now called "Sirius A"), while the *dotted line* and *open circles* represents the positions of Sirius B. The *dashed diagonal line* shows the motion of the center of mass and is the path Sirius would have followed if it were not part of a binary system. Reproduced with Holberg's permission

Observatory. In 1851, Peters systematically extended Bessel's analysis of the proper motions of Sirius and Procyon. From his new analysis, which included many additional observations and improved corrections to the data, Peters found that the so-far-unseen companion to Sirius had an orbital period of 50 years, moved in a highly elliptical orbit, and in 1841 had most recently passed its closest approach to Sirius.

At the time Bessel was announcing his discovery of an unseen companion to Sirius, on the other side of the Atlantic Ocean a Boston tinkerer named Alvan Clark (1804–1887; see Figure 1.3) began a career as a world-famous optician and telescope maker. Seven years later, he founded Alvan Clark & Sons, a firm dedicated to the production of first-class lenses and astronomical refracting telescopes, that is, telescopes made using lenses rather than mirrors.

In 1860, Dr. Frederick A. P. Barnard (1809–1889), then president of the University of Mississippi, commissioned the Clark

FIGURE 1.3 Alvan Clark. Public domain image, courtesy Wikimedia Commons (http://commons.wikimedia.org/wiki/File:1891_Alvan_Clark_Boston.png)

firm to build what was then the largest refractor in the world, with an 18 ½-in. objective lens. Work started in 1861, and the Clarks proposed to have it ready for Dr. Barnard to inspect by June of 1862. By then, the American Civil War had completely severed relations between the North and South, but as they had by that time invested so much effort into producing the massive lens, the Clarks decided to complete it anyway.

The story of the Clarks' accidental sighting of the faint companion to Sirius that Bessel had anticipated is well told by astronomer Jay Holberg[6]:

> *[In the] early evening of Friday, January 31, 1862… Alvan Clark and his son Alvan Jr. were using the opportunity of this cold but clear night to test the 18 ½-inch lens… It was common practice for the Clarks to field-test the resolving power of their lenses on double stars and to use bright blue stars to test the color correction. During such tests the lens cell was mounted at the end of a crane-like boom in the yard of their workshop. On this particular night the senior Clark was … pointing his instrument to the eastern sky, somewhat south of Orion. He had selected Sirius to color-test the lens, but was having trouble steadying the telescope. His son took over just as Sirius was clearing the roof of a nearby building, and*

[6] *Ibid.*, p. 67.

> *after a few seconds he noticed a very faint star in the glare of Sirius. "Father," he said, "Sirius has a companion." The elder Clark quickly confirmed what ... [his son] had seen. It is doubtful if either father or son were aware at the time of the significance of the faint companion they sighted that night, or of Bessel's earlier predictions of its existence ... since they left no written record of the historic discovery.... Alvan Clark, however, had previously discovered a number of double stars with his telescopes and was in the habit of reporting such discoveries both to local newspapers and to professional astronomers.*

In keeping with the designations of other binary star systems, the primary star was given the name Sirius A, while the faint companion became known as Sirius B. The Clarks' discovery revealed that Sirius B is nearly 10,000 times fainter than Sirius A, or about one 400th as bright as the Sun would be if placed at the same distance from Earth. The discovery that the mysterious companion is so much fainter than Sirius itself provided a second clue to its surprising nature.

Ever since Johannes Kepler's 1609 publication of *Harmonice Mundi*, scientists had known that the squares of the periods of planetary orbits are proportional to the cubes of the semi-major axes of their elliptical orbits (Kepler's third law). Isaac Newton's publication of his law of universal gravitation in 1687 both explained this relation and generalized it to stars as well as planets. With accurate determinations of the periods and dimensions of the orbits of binary stars, astronomers thus acquired a tool that enabled them to measure the masses of stars. For the Sirius system, Peters' 1851 determination of the orbital period is very close to the current value[7] of 50.075 ± 0.103 years, while Elkin's 1870 value for the semi-major axis is similarly close to the current value[4] of $7.50 \pm 0.03''$. With these values, Newton's generalization of Kepler's third law gives the sum of the masses of Sirius A and Sirius B as about three times the mass of our Sun.

These results for the combined masses of Sirius A and B only needed an estimate for the ratio of the masses of the two stars to obtain their individual values. This ratio had actually been obtained in 1866 by Otto Struve, the director of the Pulkovo

[7] Holberg (2007), p. 232

Observatory near St. Petersburg in Russia. He found that the observed angular separation of Sirius A and B was about three times as large as the angular separation of Sirius from the center of mass of the system, and this immediately gave the mass of Sirius B as about half of the mass of Sirius A. In other words, the mass of Sirius itself was about twice the mass of the Sun, while that of its faint companion was about equal to the solar mass. This provided a third clue to the strange nature of Sirius B.

By the end of the nineteenth century, both the orbit and the masses of Sirius A and B were thus known with reasonable accuracy. It was puzzling that Sirius B was also known to be intrinsically faint, but as the temperature of the star was then completely unknown, the faintness could be accommodated if the temperature of the faint star were low enough. However, when the first estimate of the temperature of Sirius B was obtained early in the twentieth century, astronomers experienced a profound shock.

2. A Star the Size of the Earth? Absurd!

During the final decade of the nineteenth century, Edward C. Pickering (1846–1919), the director of the Harvard College Observatory in Cambridge, Massachusetts, recruited a staff of young women for a number of routine observatory tasks (See Figure 2.1). He gave one of them—Williamina P. Fleming (1857–1911)—a tedious but extremely important assignment: searching through and classifying hundreds of thousands of tiny images of stellar spectra. As we shall see, Fleming discovered the next clue to the mysterious nature of Sirius B.

In a sense, however, this part of the story began in England more than two centuries earlier, with Isaac Newton's (1642–1727) experiments in optics. A farmer's son from the small Lincolnshire village of Woolsthorpe, Newton had traveled to Cambridge to begin his university studies in 1661. When the university was closed in 1665 because of plague in the town, Newton returned home and embarked on a program of self study that was to revolutionize mathematics, physics, and astronomy. During those remarkable years, he developed calculus, conducted fundamental experiments in optics, and advanced the laws of mechanics and the theory of universal gravitation.

In one of his experiments, Newton directed a beam of sunlight through a glass prism and discovered that it was dispersed into all the different colors of the spectrum. No doubt others had noticed this phenomenon before, but it took Newton's genius to appreciate its true importance. In a letter to the Royal Society in 1672, Newton wrote[1], "I saw that … light itself is a heterogeneous mixture of differently refrangible rays."

[1] Pannekoek (1989), p. 294.

H.M. Van Horn, *Unlocking the Secrets of White Dwarf Stars*,
Astronomers' Universe, DOI 10.1007/978-3-319-09369-7_2

FIGURE 2.1 Harvard College Observatory Director Edward C. Pickering poses with a group of women staff members on 13 May 1913. *Front row* (*left* to *right*): Margaret Harwood, Arville Walker, Johanna Mackie, Alta Carpenter, Mabel Gill, Ida Woods, Grace Brooks. *Back row*: Mollie O'Reilly, E. C. Pickering, Edith Gill, Annie J. Cannon, Evelyn Leland, Florence Cushman, Marion Whyte. Image courtesy Harvard College Observatory. Reproduced with permission

Almost a century and a half later, Joseph Fraunhofer (1781–1826), a gifted young Bavarian optician and glassworker, made the next advance in understanding the spectrum of visible light. To improve the lenses he was creating, Fraunhofer conducted detailed studies of the refraction of light of different colors by different types of glass. In the process, he discovered that the spectrum of sunlight was actually crossed by a large number of fine black lines. In an 1817 paper for the Bavarian Academy he wrote,[2] "By means of many experiments and variations, I have become convinced that these lines ... belong to the nature of solar light ... " Now called "Fraunhofer lines," these dark striations always occur at precisely the same wavelengths in the spectrum. Fraunhofer marked the strongest of these lines with the letters A through

[2] *Ibid*, p. 330.

H. By mid-century, a number of scientists had recognized that a strong double line in the solar spectrum, which Fraunhofer had labeled "D," coincided precisely in wavelength with the bright yellow double line in the spectrum of sodium. The obvious conclusion was that sodium is present in the Sun.

During the years 1859–1862, physicist Gustav Kirchoff (1824–1887) established three important "laws of radiation"[3] that provided a solid foundation for Fraunhofer's observations: (1) A hot, glowing solid or dense gas emits a continuous spectrum of radiation, without light or dark lines. (2) A hot, diffuse gas, such as a flame, produces a spectrum consisting of bright lines. (3) When viewed through an intervening cool gas, a continuous spectrum acquires dark lines at wavelengths characteristic of the chemical elements in the intervening gas.

Kirchoff's laws for the first time enabled astronomers to determine the chemical composition of a distant star. Measuring the wavelengths of some thousands of Fraunhofer lines, Kirchoff showed that they coincided with lines emitted by chemical elements such as hydrogen, iron, sodium, magnesium, calcium, etc., and he concluded that these same elements were present in the gaseous atmosphere of the Sun. A dramatic verification of Kirchoff's method of spectrum analysis was the 1868 detection of lines in the solar spectrum of a then-unknown element—helium—prior to its discovery on Earth.

Another important development in the closing decades of the nineteenth century was the experimental and theoretical effort to understand the properties of radiation in equilibrium with matter. Kirchoff showed that the spectrum of such thermal radiation (also called black body radiation) depends only upon the temperature of the matter and not on any other material properties. From measurements over a large range of temperatures, the Austrian physicist Josef Stefan (1835–1933) found in 1879 that the total energy density of such thermal radiation, summed over all wavelengths, varies in proportion to the fourth power of the absolute temperature.[4] Thus, doubling the temperature of a radiation source increases the energy density by a factor of 16. The eminent German physicist

[3] Holberg (2007), p. 86.

[4] Absolute temperature, in degrees Kelvin and abbreviated "K," is measured from 0 at approximately −273 °C. See the Glossary in this book for additional information.

Ludwig Boltzmann (1844–1906) provided a rigorous theoretical foundation for this relation in 1884, and it is now known as the "Stefan-Boltzmann law."

In 1893, Wilhelm Wien (1864–1928) demonstrated that the thermal radiation emitted from a cavity or a perfect black body in thermal equilibrium is given by a unique function of the wavelength and that the wavelength at which the radiation intensity reaches its peak value varies inversely with the temperature. In other words, if the maximum intensity in the spectrum of a source at one temperature occurs at 7,000 Å[5] (at the far red end of the spectrum), doubling the temperature shifts the peak to 3,500 Å (in the ultraviolet part of the spectrum). These two laws—the Stefan-Boltzmann law and Wien's "displacement" law—when applied to the observational data for the Sun showed that the Sun's surface temperature was nearly 6,000 K.

As soon as Kirchoff had demonstrated the value of spectrum analysis, astronomers began to attach spectroscopes to their telescopes in order to analyze the spectra of the brighter stars. This work was greatly facilitated by the application of photographic technology to astronomy beginning in the middle of the nineteenth century, especially after sensitive silver bromide-gelatin photographic plates were introduced in 1871.

Henry Draper (1837–1882) in particular focused on this work. A medical doctor and dedicated amateur astronomer in New York City, he had ground a mirror 72 cm in diameter (approximately 28 in.) for his telescope, and by placing a quartz prism in front of the focus, he was able to disperse starlight into a spectrum. In 1872 he succeeded for the first time in photographing the spectrum of the bright star Vega. Continuing this work in 1879 with a 28-cm (11-in.) refracting telescope and a Browning spectrograph, he was able to photograph the spectra of some 50 stars. When Henry Draper died in 1882, his widow donated his instruments to the Harvard College Observatory, together with a sum of money for a "Henry Draper Memorial Fund" to enable Draper's work on stellar spectra to be continued.

As director of the Harvard College Observatory at the time of the Draper bequest, Edward Pickering used the money to equip a

[5] One Ångstrom unit, abbreviated "Å," is ten billionths of a cm, or 10^{-8} cm.

wide-angle telescope with an objective prism—a large glass disk ground to be slightly thinner on one side than on the other, creating a large, flat, round prism with a small refractive angle. When the disk was inserted into the telescope, the images of stars on the focal plane became small spectra, instead of small, round points. A single photographic plate could thus contain the tiny spectra of all the hundreds of stars in a large field at the same time.

This brings us up to the point in our story at which Professor Pickering assigned Williamina Fleming the job of classifying the large number of stellar spectra. With no formal astronomical training or preconceptions, she simply grouped bright blue stars like Sirius together in a class she labeled "A." Another group of spectra that were similar to each other she labeled "B," and so on up through the alphabet. Stars like the Sun she termed spectral class G, and she grouped together as spectral class M a group of stars with reddish spectra and strong, dark bands (which later proved to be produced by molecules formed in the atmospheres of these cool stars), in addition to the narrow, dark lines due to individual atoms. After the element helium was discovered in the solar spectrum, the further spectral class O—typified by the presence of lines due to singly ionized helium—was added; the hot, bright blue stars in the constellation Orion belong to this class. The results, published in 1890 in the first *Henry Draper Catalogue*, gave the spectral classifications for some 10,000 stars.

In 1896, Pickering added Annie Jump Cannon (1863–1941) to the staff at the Harvard College Observatory and assigned her the task of examining some peculiar spectra found in the first Draper survey. Cannon came to this work with some graduate-level training in astronomy, and she soon developed a special facility in distinguishing minute differences in the stellar spectra. As a consequence, she was able to simplify the older classification system considerably, retaining only the spectral classes O, B, A, F, G, K, and M, while adding decimal subdivisions to accommodate intermediate spectral types. In Cannon's revised system, for example, the Sun is classified as belonging to spectral class G2.

From the changes in the state of ionization and excitation of the elements identified in the stellar spectra, it was generally recognized that the spectral classes represented a relative temperature sequence, with the blue O stars (with lines due to ionized helium) at the hottest end, the B stars (with neutral helium lines) and A stars (with neutral hydrogen lines) next, the yellow F and G stars (including the exact solar spectrum) at intermediate temperatures, and the reddish K and M stars (with molecular bands) at the coolest temperatures. In the current spectral classification system, the spectral types running in order from the hottest to the coolest stars have become known to generations of astronomy undergraduates through the mnemonic "Oh Be A Fine Girl/Guy, Kiss Me." After many years of work, the revised and extended *Henry Draper Catalogue,* containing the magnitudes and spectra of 225,000 stars, was completed and published during period of 1918–1924 in nine volumes of the *Harvard Annals.*

It was against this background of the development of a rigorous system of spectral classification that astronomers in 1910 finally recognized that white dwarfs were a completely different, new class of stars. The key discovery involved the tenth magnitude star 40 Eridani B, in the constellation Eridanus. With its primary, 40 Eridani A, a fourth magnitude K star, it forms a binary with a period of 200 years. Having a large proper motion,[6] the system was known to be relatively nearby. This was confirmed by its large parallax—0.20″—which together with its apparent magnitude of +10 made 40 Eridani B an intrinsically faint star, with a luminosity of about 100 times less than its primary, or about 400 times smaller than that of the Sun.

The orbital motions of the two stars yielded for 40 Eridani B a mass of 0.4 times the mass of the Sun. Up to this point, its characteristics were consistent with it being a member of the class of red dwarf stars of spectral type M, which abounded in the *Henry Draper Catalogue.* That changed dramatically in 1910, however, when astronomer Henry Norris Russell (1877–1957) asked to have the spectral type of 40 Eridani B determined. The story, as recalled by Russell in the 1940s, is quoted in Holberg's book[7]:

> *The first person who knew of the existence of white dwarfs was Mrs. [Williamina] Fleming; the next two, an hour or so later, Professor E. C. Pickering and I. With characteristic generosity,*

[6] "Proper motion" is the apparent distance that a star moves across the sky in 1 year. A star with a large proper motion is generally nearer to Earth than one with a small proper motion. See the Glossary in this book for additional information.

[7] Holberg (2007), p. 114.

Pickering had volunteered to have spectra of the stars which I had observed for parallax looked up on the Harvard [photographic] plates. All those of faint absolute magnitude turned out to be of class G or later [i.e., cooler]. Moved with curiosity I asked him about the companion of 40 Eridani. Characteristically, again he telephoned to Mrs. Fleming, who reported within an hour or so that it was of Class A. I saw enough of the physical implications of this to be puzzled ...

Three years later, the Dutch astronomer Adriaan van Maanen (1884–1946) found that the faint, single star van Maanen 2, which he had discovered, had a similarly incongruous "hot" spectral type, and in 1915, the Danish astronomer Ejnar Hertzsprung (1873–1969) confirmed that 40 Eridani B was of spectral type A; that is, it was hot enough to show hydrogen lines in its spectrum. Also in 1915, Walter S. Adams (1876–1956) succeeded in photographing the spectrum of Sirius B using the 60-in. telescope at Mt. Wilson. This was an exceptionally difficult observation, because Sirius A is so much brighter than Sirius B, and the two stars are close together in the sky. Adams succeeded in 1915 because Sirius B was then near its maximum separation from Sirius A in its orbit. When Adams developed the photograph, he found that the spectrum of Sirius B contained nothing but hydrogen lines. Since Sirius A also contains hydrogen lines, he concluded that the two stars had to have similar temperatures. These three stars established the existence of a then-new class of astronomical objects. Because of their white color and intrinsic faintness (and hence small size), such stars were named "white dwarfs."

The discovery that the temperatures of Sirius A and B were surprisingly similar immediately enabled astronomers to infer the size of Sirius B. Since the luminosity of a star is proportional to the square of its radius (i.e., to its surface area) multiplied by the fourth power of its surface temperature (from the Stefan-Boltzmann law), then if the temperatures were essentially the same the squares of the radii of the two stars must be proportional to their respective luminosities. As the luminosity of Sirius B had been known to be about 10,000 times fainter than that of Sirius A ever since the Clarks had first spotted the faint companion more than half a century earlier, the inescapable conclusion was that Sirius B was only about a hundredth the size of Sirius A. That is, the radius of Sirius

B was roughly 13,000 km, or about twice the size of Earth. (Today, we know that Sirius B actually is almost three times hotter than Sirius A, so that the radius is correspondingly smaller still, but that is a later part of the story.) Incredibly, Sirius B packed a mass as large as the Sun's into a volume about the size of Earth, which the scientists of the day considered absurd.[8]

The discovery that the size of Sirius B was so small, coupled with its known, substantial mass, had some immediate and unsettling consequences.

First, the force of gravity at the surface of the star was enormously greater than the surface gravity of any other astronomical body then known. (It has since been exceeded by the enormous surface gravities of neutron stars.) Where the surface gravity of our Sun is about 30 times larger than that of Earth—so that an object weighing 100 lb on Earth would weigh about a ton and a half at the surface of the Sun—the surface gravity of Sirius B is 10,000 times larger still. The same object, if placed on the surface of Sirius B, would thus weigh more than 18,000 tons, as much as two modern guided missile destroyers weigh on Earth!

A second consequence of the large mass and small radius of Sirius B is that the *average* density of the star (mass divided by volume) approaches a million grams per cubic centimeter. That is, a chunk of white-dwarf stuff that is about the size of a sugar cube contains as much mass as a small car. And that is just the average density; at the center of the star the density is many times larger still! Such high densities carried profound implications about the properties of white-dwarf matter that could not begin to be understood until quantum physics was developed over the next few decades.

[8] Eddington (1926), p. 171.

3. Great Balls of Fire

By the latter part of the nineteenth century, astronomers had learned a considerable amount about the nature of the stars. From accurate measurements of stellar parallaxes, they had determined the distances to the nearest stars; if the Sun were removed to such a distance, it would appear as a very ordinary star, showing that the Sun and the stars are similar objects. The laws of gravitation and mechanical motion, developed through terrestrial experiments and then extended to the planets in the Solar System, had been applied to great advantage in analyzing systems of binary stars. The development of spectrum analysis had shown that the Sun and stars were made of the same elements as found on Earth. And the discovery that the surface temperature of the Sun was very high implied that the elements in the Sun all had to be in gaseous form.

A first step toward the development of a physical model for the internal structures of the stars came in 1862, when William Thomson (Lord Kelvin, 1824–1907) published a theory describing the way heat is transported by convective motions in a fluid (or gas) under gravity, such as in the interior of a star.[1] Kelvin realized that the process depends critically upon the gradient of temperature across the fluid.

Suppose a parcel of fluid is displaced vertically upward from its original location. Assuming the fluid to be in hydrostatic equilibrium under gravity, the ambient pressure at the new (higher) position is less than that at the initial (lower) location, so the fluid element must expand until the pressures inside and outside are equal. If the expansion preserves the initial heat content of the fluid element—a process termed an "adiabatic" expansion—then the temperature and density within the fluid element will in general differ from the external pressure and density at the new

[1] Chandrasekhar (1939), pp. 84–85.

H.M. Van Horn, *Unlocking the Secrets of White Dwarf Stars*,
Astronomers' Universe, DOI 10.1007/978-3-319-09369-7_3

location. If the internal density is less than the ambient external density, then the parcel is buoyant and will continue to rise—that is, convective motions will arise spontaneously. This occurs when the gradient of the temperature in the ambient fluid is steeper than would be the case for an element displaced adiabatically.

Conversely, if the density inside the fluid element exceeds the external density—or equivalently if the gradient of the external temperature is less steep than the adiabatic gradient—the fluid parcel is denser than its surroundings and will sink back to its original location—that is, the fluid is stable against convection. When convection does occur, it turns out to be so efficient in transporting heat that the actual temperature gradient is only slightly steeper than the adiabatic gradient. Such a region is said to be "in convective equilibrium."

In 1869, an American applied mathematician named J. Homer Lane (1818–1890) took the next step, making the first attempt to develop a physical model for the Sun.[2] Although he was primarily interested in the temperature and density at the solar surface, he also modeled the Sun as a gaseous sphere in mechanical equilibrium, with gas properties as determined in terrestrial laboratories. Robert Boyle (1627–1691) had shown in 1660 that the pressure of an ideal gas varies inversely with its volume (or, equivalently, in proportion to its density), and in 1802 Joseph Louis Gay-Lussac (1778–1850) added that, for a fixed density, the pressure varies in proportion to the temperature.

Lacking a way to determine the temperature inside the Sun, Lane instead assumed it to be in convective equilibrium. From Lane's viewpoint, this provided the simplifying feature that the pressure at any point within the interior was proportional to a known, fixed power of the local density (the adiabatic pressure-density relation) and did not depend separately upon the temperature. With the well-justified assumption that the Sun is in hydrostatic equilibrium, Lane was thus able to write two equations to describe the internal structure of the Sun, one expressing conservation of mass and the second expressing the condition of hydrostatic equilibrium, which he solved numerically.

[2] *Ibid.*, p. 176; see also Holberg (2007), p. 108.

A decade later, the German astrophysicist August Ritter (1826–1908) published a series of technical papers in which he extended this type of modeling to the stars.[3] In the process, he laid the foundation for much of the mathematical theory of stellar structure subsequently developed. In 1887, Lord Kelvin also took up this problem. All three authors (Lane, Ritter, and Kelvin) made the simplifying assumption that the pressure at any point in the star was proportional to a power of the local mass density. This general type of pressure-density relation is termed a "polytropic equation of state," and the adiabatic pressure-density relation is a special case. The resulting equations describing the structure of a stellar model require a numerical solution.

In his classic 1939 treatise on stellar structure, Subramanyan Chandrasekhar (1910–1995) later provided a thorough mathematical exposition of the solutions of these equations.[4] Such polytropic models have a number of properties that helped early investigators make significant advances in understanding the interiors of real stars. For example, given values for the exponent and the constant of proportionality in the pressure-density relation, there is a unique relation between the mass and radius of the model and a unique value for the ratio of the central density to the mean density. There are also unique expressions for the pressure at the center of the model, for the gravitational potential energy, and for the total internal energy.

Despite the overly simplified nature of these polytropic models, they provided the first real insights into physical conditions inside stars. For example, if we consider a fully convective model, the central density is approximately six times larger than the mean density. If the model is taken to have the same mass and radius as the Sun, the mean density is 1.4 g cm^{-3}, and the central density is about 8.4 g cm^{-3}. In addition, the central pressure of the model is approximately 9×10^{15} dyn cm^2, or about 9 billion atmospheres.[5] For an ideal hydrogen gas at this same pressure and density, the temperature is about 12 million degrees Kelvin (12×10^6 K).

[3] Chandrasekhar (1939), p. 177; see also Pannekoek (1989), p. 417.

[4] Chandrasekhar (1939), p. 84 ff.

[5] See Appendix B in this book for more about powers of ten and logarithms.

For comparison, a more recent, detailed computer model for the interior of the present Sun has a central density of 158 g cm^{-3}—almost 10 times higher than the density of gold under terrestrial conditions!—and a central temperature of 15.7×10^6 K. Thus, even though the Sun is *not* a fully convective star consisting solely of hydrogen, the polytropic value for the central temperature was close enough to the true value at the center of the Sun to help guide subsequent thinking about stellar interiors.

While some theorists were advancing the understanding of stellar interiors, other astronomers were finding important correlations among observed stellar properties. During the first decade of the twentieth century, Danish amateur astronomer Ejnar Hertzsprung (1873–1967; see Figure 3.1) published a number of important papers on the intrinsic luminosities of the stars.

Realizing that all stars in clusters such as the Pleiades lie at about the same distance from Earth, he recognized that the relative differences in apparent magnitudes for the cluster stars reflected the differences in their absolute magnitudes (the apparent magnitude if the star were placed at a standard distance of 10 pc). The absolute magnitude is proportional to the logarithm of the stellar luminosity (total power output over all wavelengths). When Hertzsprung plotted the absolute magnitudes of the cluster

FIGURE 3.1 Ejnar Hertzsprung. Credit: Novdisk Pressefoto A/S, courtesy AIP Emilio Segré Visual Archives, gift of Kaj Aage Strand

stars against crude measures of their colors[6]—which in turn depend upon their surface temperatures—he found that most stars fell along a diagonal band that he termed the "Main Sequence," a name we use to this day. He also found that a few of the stars in the cluster were very luminous red stars.

About this same time, Henry Norris Russell (1877–1957; see Figure 3.2) was independently working on the problem of stellar luminosities at Princeton University, where he had become a junior faculty member in 1905. Russell approached the problem by using parallax measurements he and others had made, as well as a variety of innovative indirect methods, to determine the distances to individual stars. He formed an early friendship with Edward C. Pickering at the Harvard College Observatory, and Pickering provided him with determinations of the spectral classes that Russell had obtained distances for.

FIGURTE 3.2 Henry Norris Russell. Public domain image courtesy Wikimedia Commons (http://upload.wikimedia.org/wikipedia/commons/a/af/Portrait_of_Henry_Norris_Russell.jpg)

[6] Chandrasekhar (1939), p. 291; see also Holberg (2007), pp. 101 ff.

In 1913, by plotting the stellar luminosities derived from the distance measurements Russell had acquired against the Harvard spectral classes, he found[7] essentially the same type of diagonal "Main Sequence" band that Hertzsprung had found, as well as a number of stars that lay above the Main Sequence, at redder colors (and thus lower surface temperatures) and higher luminosities (see Figure 3.3). Since stars with higher luminosities at the same or lower temperatures must have larger radii (from the Stefan-Boltzmann law), Russell called them "giants," while the Main Sequence stars he termed "dwarfs," erroneously attributing the names to Hertzsprung. The few white dwarf stars known by the middle of the second decade of the twentieth century occupied a lonely position by themselves at lower luminosities than any other stars with similar colors or spectra. The Hertzsprung-Russell diagram has since become an important tool for astronomers, providing a map of the relationships between different phases of stellar evolution, as we shall see in Chap. 10.

While these developments were taking place in astronomy, a revolution was brewing in physics that would prove to have major consequences for the understanding of the stars. The first hint of this came in 1885, when physicist Johann Jacob Balmer (1825–1898), working at Basel in Switzerland, found that the strong lines of hydrogen in the visible part of the spectrum followed a simple relation in which the wavelength of each hydrogen line was given by a limiting value of 3,645.6 Å multiplied by the expression $1/[1-(2/n)^2]$, where n is any integer greater than 2. Other scientists soon discovered other series of lines in the hydrogen spectrum, each following different but similar progressions in wavelength. As astronomers were to learn, most of the matter in ordinary stars such as the Sun consisted of hydrogen, so the "Balmer lines" of hydrogen (as they came to be called) were to play an important role in the spectral classification of stars. Balmer's formula was also to play a key role in the establishment of quantum theory in the early decades of the twentieth century.

The next piece of the story was provided in 1900 by German physicist Max Planck (1858–1947), who was trying to calculate theoretically the spectrum of thermal or "black body" radiation.

[7] Holberg (2007), pp. 103 ff.

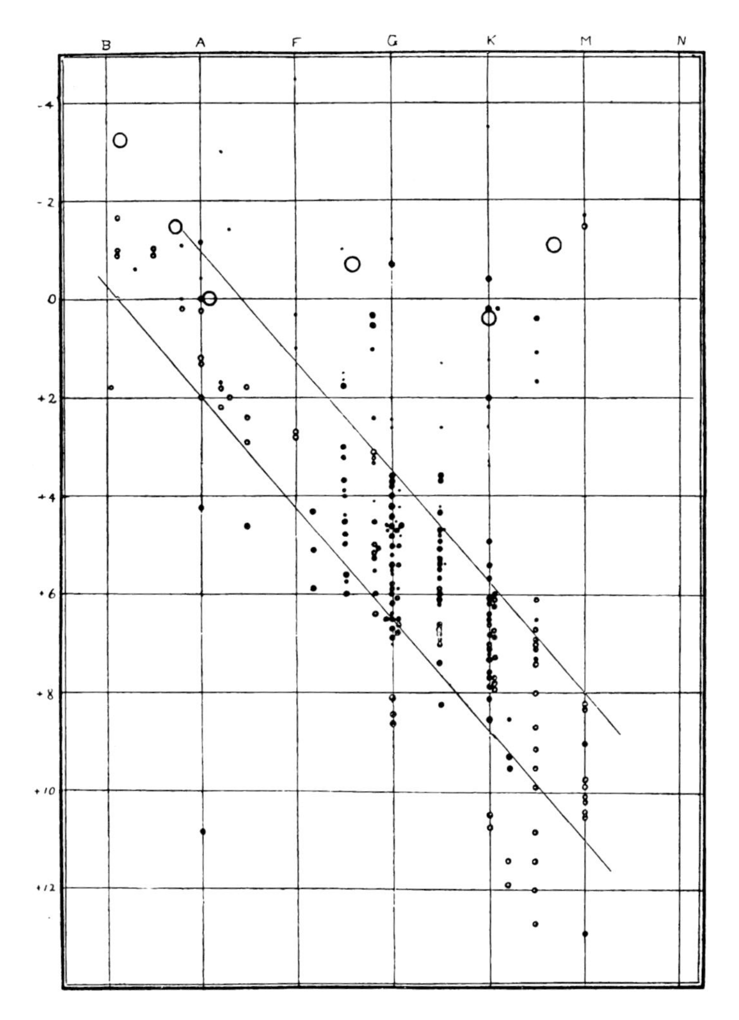

Figure 3.3 Russell's 1914 version of the Hertzsprung-Russell diagram. Reprinted by permission from Macmillan Publishers Ltd: *Nature*, **93**, 252, copyright 1914. Points represent individual stars, which are plotted by spectral type (ranked left-to-right from hottest to coolest) and by absolute magnitude (from brightest at the *top* to faintest at the *bottom*). The Main Sequence is represented by the stars between the two *diagonal lines*, while red giants populate the *upper right-hand* part of the diagram. The lone white dwarf, 40 Eridani B, is the single, isolated point in the *lower left* part of the figure

To get rid of some problems in the calculations, Planck made the radical assumption that exchanges of energy between matter and radiation take place not in a continuous fashion, as physicists then assumed, but through emission or absorption of distinct quanta of radiation (now termed "photons").[8] He postulated that a photon of frequency ν carries a corresponding amount of energy $h\nu$, where h is a constant that remained to be determined. This has since come to be called "Planck's constant" and has the value 6.625×10^{-27} erg s^{-1}, where an erg is a unit of energy. With this radical hypothesis, Planck found that he was in fact able to calculate exactly the experimentally measured shape of the spectrum of thermal radiation.

At the time, most physicists ignored Planck's result, regarding it as nothing more than a mathematical trick, with no basis in reality. This changed in 1905, however, when the young Albert Einstein (1879–1955) showed in one of three remarkable papers that if one took Planck's idea seriously it was possible to explain the otherwise puzzling photoelectric effect, in which the energies of electrons ejected when light was shined on a metal surface depended upon the frequency of the light but not on its intensity.

In 1913, Danish physicist Niels Bohr (1885–1962) provided the explanation for the mysterious pattern of lines Balmer had found in the spectrum of hydrogen.[9] Two years earlier, British physicist Ernest Rutherford (1871–1937) had shown experimentally that an atom consisted of a small, positively charged nucleus around which lighter, negatively charged electrons orbit like planets in a miniature Solar System. With the electrostatic force between the negative electrons and the positive nucleus providing the force of attraction binding the atom together, Bohr proposed that the electrons could only occupy certain discrete orbits—those in which the orbital angular momentum of the electron was proportional to an integer multiple of Planck's constant h.

When radiation is either emitted or absorbed by the atom, Bohr continued, the electron jumps from one discrete orbit to another. The energy of the emitted or absorbed radiation thus occurs in discrete quanta, just as had been postulated by Planck. Bohr's model provided a direct explanation for the existence of spectral lines, and the energies he calculated for transitions in the

---

[8] Planck's classic paper containing his quantum hypothesis has been republished, both in the original German and in English translation, by Kangro (1972). It is also discussed by French (1958), p. 92 ff. See also Pannekoek (1989), p. 414, and Langer (1968), p. 597.

[9] French (1958), p. 107.

hydrogen atom between the second allowed electron orbit and orbits with higher energies followed exactly the pattern of spectral lines observed in the Balmer series. And since the energies of the higher levels crowded together more and more closely, the model also provided the experimentally determined value for the limit of the Balmer series. If a photon of sufficiently high energy illuminated the atom, the electron would be ejected, leaving behind a positively charged ion. Bohr's model, which unified Planck's quantum hypothesis for radiation with a quantized model for the atom, eventually came to be called the "old quantum theory," after it was replaced a decade or so later by the "new quantum theory."

The development of quantum theory enabled physicists to calculate the absorption of radiation by atoms and ions, and this provided the crucial next step in the advancement of stellar models. The main shortcoming of the early polytropic models for stellar interiors had been the approximate treatment of the local temperature. To improve the models would require adding an equation for the temperature gradient in the stellar interior. This in turn necessitated treating the transport of energy by radiation. For such radiative energy transport, a quantity termed the "opacity"—which depends upon the ability of matter to absorb radiation—determines the efficiency of the process. Matter with a high opacity traps the radiation, allowing it to leak through only slowly, while for matter with a low opacity radiation can flow through relatively easily.

The first step in incorporating heat transport by radiation was taken in 1905 by the brilliant German astrophysicist Karl Schwarzschild (1873–1916), who established the concept of radiative equilibrium and applied it to the solar atmosphere.[10] Schwarzschild made a number of other profound contributions to physics and astrophysics before his life was cut short during World War I.

In contrast to the case of convective equilibrium considered earlier, the condition of radiative equilibrium postulates that all of the energy is instead carried by radiation. In effect, photons undergo a "drunkard's walk" through the region they are traversing, continually being absorbed and then re-emitted in some random direction, gradually diffusing from a region that provides the

[10]Chandrasekhar (1939), p. 183 ff.

FIGURE 3.4 Arthur Stanley Eddington. Credit: AIP Emilio Segré Visual Archives, gift of Subrahmanyan Chandrasekhar

source of the photons, through the intervening matter, and finally being radiated out into space.

In 1916, British astrophysicist Arthur Stanley Eddington (1882–1944; Figure 3.4), one of the leading scientists of his day, extended Schwarzschild's work on radiative equilibrium to the interiors of the stars. He included several major improvements over the earlier polytropic models.

First, he incorporated the pressure produced by the great intensity of the radiation inside the star in balancing gravity to establish hydrostatic equilibrium. With temperatures near the stellar center exceeding ten million degrees Kelvin, almost all atoms are completely ionized, and the main contributions to the opacity are provided by the scattering of radiation by free electrons and the absorption of energy by an unbound electron, taking it from a quantum state with lower energy to one with higher energy (called "free-free absorption"). At the somewhat lower temperatures occurring farther out in the star, an increasing fraction of the atoms—especially those like iron with more highly charged nuclei that require higher energies to ionize—retain some bound electron states. This produces a further contribution to the opacity when the absorption of a photon ejects a bound electron (called "bound-free absorption").

Eddington incorporated approximations for these opacity sources in computing the transport of radiation from the million-degree temperatures of the deep interior to the surface. Eddington also assumed that the source of energy to maintain the flow of radiant energy had to be concentrated at the center of the star. The physical nature of this energy was still unknown at the time of Eddington's work, but these improvements enabled him to determine with greatly increased accuracy the temperature, pressure, density, degree of ionization, and opacity as functions of distance from the center of the stellar model. Eddington summarized a decade of work on the physical properties of stellar interiors in his classic 1926 text *The Internal Constitution of the Stars*.[11]

A dozen years later, the Indian astrophysicist Subramanyan Chandrasekhar placed the capstone on this second epoch in the study of stellar structure with his authoritative text *An Introduction to the Study of Stellar Structure*. He provided a thorough exposition of the physical principles governing stellar interiors, beginning with thermodynamics, progressing to theorems concerning a star in equilibrium, and continuing through detailed polytropic models. He next considered the emission and absorption of radiation by elements of matter, calculated the energy density and pressure associated with radiation, and derived the equation governing the transport of radiation through the star. He continued with a discussion of the outer layers ("envelopes") of stars, and developed a framework for modeling stellar interiors.

Chandrasekhar and others showed that the structure of a star is determined by four equations: one governs conservation of mass; a second governs hydrostatic equilibrium; a third governs energy transport through the star by radiation, convection,

[11] Eddington's work on modeling the interiors of stars is discussed by Holberg (2007), pp. 109 ff; see also Pannekoek (1989), pp. 417–418, 464–465.

or conduction; and the fourth governs conservation of energy. In the last chapters of his book, Chandrasekhar considered quantum statistics and the theory of white dwarf stars (to which we shall turn in the next chapter), and he ended with a consideration of stellar energy sources, including a brief discussion of the possibility of nuclear energy sources.

4. Relativity, Wave-Particle Duality, and the Nature of White Dwarfs

The early decades of the twentieth century saw two revolutionary advances in physics that were to prove essential to understanding the nature of white dwarf stars.

The first was due to Albert Einstein (see Figure 4.1), then an obscure "technical expert, third class," working in the patent office in Bern, Switzerland. Born in Ulm, Germany, the young Albert moved to Zürich, Switzerland, in 1896 and became a naturalized citizen so that he could attend the Swiss Federal Technical Institute. He pursued a degree in physics, mostly through self-directed study, as he apparently had little patience with his professors' approach to teaching. Finishing his studies in 1900, he qualified as a teacher of mathematics and physics but was unable to find a university position. After a number of temporary teaching positions, in 1902 he found a job at the Swiss patent office in Bern. This was apparently more to his liking, as it gave him time to pursue his own investigations.

Einstein's efforts very soon became astonishingly fruitful, and in the "miraculous year" 1905, he published three extraordinary papers.[1] One, explaining the previously mysterious photoelectric effect as a consequence of the absorption of a quantum of radiation (or "photon"), pre-dated Bohr's model for the atom. A second paper explained the puzzling phenomenon called "Brownian motion"—the erratic motion of microscopic particles in a liquid—as a consequence of the jostling of the particles by the random thermal motions of atoms in the liquid.

Einstein's third paper, which completely transformed the science of physics and in the process provided one of the key pieces to

[1] Pais (1982), pp. 17–21.

H.M. Van Horn, *Unlocking the Secrets of White Dwarf Stars*, Astronomers' Universe, DOI 10.1007/978-3-319-09369-7_4

FIGURE 4.1 Albert Einstein in the patent office in Bern, Switzerland, about 1905. Credit: Photograph by Lucien Chavan, Hebrew University of Jerusalem Albert Einstein Archives, courtesy AIP Emilio Segré Visual Archives

the puzzle of white dwarf stars, described the special theory of relativity. Although Einstein's self-confidence in his understanding of physics was apparent in all three papers, his ability to recognize flaws in reasoning to which others were blind and his genius in following a rigorous argument to its logical conclusion were vividly displayed in this third paper. He made two assumptions: (1) that the laws of physics take the same form in any reference frame moving with a constant velocity and (2) that the speed of light has the same value no matter what the velocity of the body from which the light is emitted. These two assumptions enabled him to construct a new, self-consistent theory of bodies in uniform relative motion—the special theory of relativity—which had amazing consequences.

The first consequence was that time had to be regarded as a fourth dimension, complementing the usual three spatial dimensions. The second was the discovery that mass and energy are intimately related, as summarized in his famous equation $E = mc^2$, where E represents energy, m stands for the mass of a body at rest, and c denotes the speed of light. Within a few decades, this equation would lead to a fundamental understanding of the energy sources of the stars and unlock an awesome new source of power of terrifying magnitude. The special theory of relativity modified the relation between the energy E of a particle and its momentum, represented by the symbol p,[2] and this change provided the next key to understanding white dwarfs.

Not long after completing the special theory of relativity, Einstein began to ponder the generalization of these concepts to accelerated reference frames, rather than just to the special case of reference frames moving with constant velocities, as he realized that this would be necessary to deal with reference frames accelerating under the influence of gravitation. As he put it in 1907,[3] "... the experimentally known matter independence of the acceleration of fall is therefore a powerful argument for the fact that the relativity postulate has to be extended to coordinate systems which, relative to each other, are in non-uniform motion."

For the next several years, Einstein was preoccupied with developments in quantum theory, which was then occupying many of the best minds in physics. He returned to gravitation again in 1911 and struggled for a few years more to find a way to express the new approach he was developing. Ultimately, he concluded that what was involved was nothing less than the geometry of space itself, and by 1915 he had developed tensor equations that expressed the curvature of the four-dimensional "space-time continuum" in terms of the mass and energy contained within it.

Einstein himself proposed three tests of his new theory of gravitation, which he called the general theory of relativity. The first was the precession of the perihelion of Mercury's elliptical orbit about the Sun. Although Mercury's orbit was already known to precess, the measured value of the rate of precession differed from that given by pre-relativistic theory. The second test was the

---

[2] Pais (1982), pp. 111–134; see also French (1958), pp. 137 ff and especially pp. 149 ff; Holberg (2007), p. 142.

[3] Pais (1982), p. 178.

bending of light by the Sun, and the third was the gravitational redshift[4] of light emitted from the surface of a dense body.

All three tests were subsequently carried out, and general relativity passed them all with flying colors. Because general relativity involves the time dimension as well as space, the theory also predicts a fundamentally new phenomenon: the existence of gravitational waves. The emission of gravitational waves was verified in the 1970s by observations[5] of the binary pulsar PSR 1913+16, the gradual decay of the orbit caused by energy loss through gravitational radiation proceeding exactly as predicted by Einstein's theory.

Let us return at this point to the development of quantum physics. A flurry of publications in the mid-1920s marked the transition from the "old quantum theory" to the "new quantum theory."

One of these papers was French physicist Louis de Broglie's (1892–1989) 1924 theory that—just as light had been found sometimes to display wave-like properties (as in diffraction experiments) and sometimes particle-like properties (as in the photoelectric effect)—so, too, might material particles like electrons sometimes exhibit wave-like properties. If this were the case, de Broglie reasoned, then the wavelength of the matter waves should be given by $\lambda = h/p$, where p is the momentum of the particle and h is Planck's constant. For example, if we consider the momentum of an electron in the lowest level (called the "ground state") of Bohr's model for the hydrogen atom, then the corresponding de Broglie wavelength turns out to be about the size of the lowest Bohr orbit, or about ½ Ångstrom. Within a few years, experimental evidence for the diffraction of electrons by crystals confirmed de Broglie's hypothesis.

In 1926, Austrian physicist Erwin Schrödinger (1887–1961) developed an equation to describe the properties of such matter waves. He found that the waves representing the quantum states of an electron in a hydrogen atom are characterized by three independent

[4] We discuss the gravitational redshift of radiation emitted from the surfaces of white dwarfs in Chap. 15.

[5] The existence of gravitational waves as verified by observations of the binary pulsar PSR 1913+16 is discussed by Shapiro and Teukolsky (1983), p. 479.

"quantum numbers." The "principal quantum number," n, corresponds to the variation of the electron wave with radial distance from the proton. The second quantum number, l, corresponds to the angular momentum of the electron; for a given integer value of n, it proved to be restricted to integer values from 0 to n–1. The azimuthal variation of the electron wave is in turn specified by a third quantum number, m, which can take on integer values between $-l$ and $+l$. A fourth quantum number, s—introduced in 1925 by physicists George Uhlenbeck (1900–1988) and Samuel Goudsmit (1902–1978) to account for the observed doubling of a number of spectral lines like the sodium D lines—turned out to correspond to the intrinsic "spin" angular momentum of the electron (analogous to the rotation of Earth on its axis); it was assigned the value $s = ½$ and corresponds to two allowed orientations of the spin, generally called "up" or "down." However, only in the microscopic world of atoms and molecules are quantum effects noticeable. On a macroscopic scale, with "particles" such as baseballs or planets, quantum effects are completely imperceptible.

Following the success of quantum physics in describing the hydrogen atom, it was natural to extend quantum ideas to atoms with larger charges in the atomic nucleus (called the "atomic number" because it is equal to the number of orbiting electrons necessary to counter the positive charge and produce a neutral atom). However, an immediate problem was that all the negatively charged electrons contribute to the electrostatic potential energy felt by any one of them, unlike the hydrogen atom, in which the single electron feels only the potential produced by the positively charged proton that comprises its nucleus. The general concept of allowed quantum states for the electron, each corresponding to different values of the quantum numbers n, l, m, and s was still valid, but the calculation of electron energies for the different states became much more complicated and could only be obtained by numerical methods. This concept enabled a simple interpretation[6] of the Periodic Table of the Elements, however, and it also led to another important physical principle. Enunciated by Wolfgang Pauli (1900–1958) in 1925, the "exclusion principle" states

[6] Pauli's exclusion principle and the interpretation of the periodic table of the elements is discussed by French (1958), pp. 215–217, 225; see also Holberg (2007), p. 124.

that no two electrons can occupy states with exactly the same quantum numbers.

Another key development in the formulation of the "new" quantum mechanics was the 1927 demonstration by German physicist Werner Karl Heisenberg (1901–1976) that it is impossible to determine simultaneously and with great precision both the position and momentum of a quantum particle. According to Heisenberg's "uncertainty principle," the product of the uncertainty in the measurement of the position and the uncertainty in the momentum always exceeds Planck's constant[7] divided by 2π.

What does all this have to do with white dwarfs? To answer this question, let us consider the mean density of the white dwarf Sirius B. Using modern values for its mass and radius, the mean density (mass divided by volume) is 2.4×10^6 g cm^{-3}. This means that, on average, matter in the star is packed together so tightly that a single atom would have to occupy a tiny volume less than 5 % of the size of a hydrogen atom.

This immediately tells us two important things about matter inside a white dwarf: (1) Because this space is so much smaller than the size of an atom, the electrons in a white dwarf cannot remain bound to individual atomic nuclei. Instead, because of the high density—and correspondingly high pressure—the atoms are fully ionized (they are said to be "pressure-ionized," as distinct from the ionization produced by high temperatures), and the electrons are free to move throughout a parcel of white-dwarf matter, like electrons in a metal. (2) Because of the uncertainty principle, if an electron is localized to such a small volume, its momentum must be very high. Indeed, the speed of an electron at the average density inside Sirius B is close to the speed of light! Consequently, the effects of special relativity must be taken into account in dealing with the properties of matter inside a white dwarf.

The first person to recognize that the electrons in white dwarf matter had to be treated by taking into account their quantum mechanical properties was the British mathematical physicist Ralph H. Fowler (1889–1944).[8] In 1926, Italian physicist Enrico

---

[7] The quantity $h/2\pi$ occurs so often in quantum physics that it has been given its own special symbol, $\hbar = h/2\pi$.

[8] Holberg (2007), pp. 123 ff.

Fermi (1901–1954) and British physicist Paul A. M. Dirac (1902–1984) had developed a statistical theory for particles like electrons that obey the Pauli Exclusion Principle. They showed that when the quantum state with the lowest momentum is filled with electrons, additional electrons can only be added into the state with the next largest momentum (or energy), then on to the next quantum state, and so on until all the electrons are accounted for. Fermi and Dirac calculated the distribution of electrons in the allowable momentum states at different temperatures. They showed that at zero temperature the electrons fill up the available states to some calculable level that we now call the "Fermi energy" of the distribution, corresponding to the "Fermi momentum" of the electrons in the highest occupied level; such a distribution is said to be "degenerate."

Aware of the work of Fermi and Dirac, Fowler realized that the same type of calculation could be applied to the matter inside a white dwarf. If the electrons really were moving at nearly the speed of light, then the electrostatic interactions between the electrons and nuclei would be much smaller than the kinetic energies of the rapidly moving electrons, so that in a first approximation it was reasonable to neglect the electrostatic interactions entirely. Furthermore, such speeds were so much larger than the thermal velocities of electrons at any conceivable temperature that it was also reasonable to approximate the calculation by taking the temperature to be zero. And since particles obeying Fermi-Dirac statistics have a distribution that extends up to large momenta at the Fermi level, they must also exert a substantial pressure—which is determined by the particle momenta—even at zero temperature.

Working out these ideas for non-relativistic electrons, Fowler calculated the pressure of such a gas of non-interacting free electrons at zero temperature and showed that it was sufficient to support an idealized zero-temperature star against gravity. The resulting models had radii and masses that were roughly in agreement with those of the few white dwarfs then known.

The obvious next step was to include the effects of special relativity. In the late 1920s a couple of calculations were undertaken to make relativistic corrections to Fowler's equation of state, but the fully relativistic equation of state and the corresponding white dwarf models were first computed by Subramanyan

FIGURE 4.2 Subramanyan Chandrasekhar, about 1939. Credit: AIP Emilio Segré Visual Archives, Physics Today Collection

Chandrasekhar (see Figure 4.2), then a brilliant young student from India.[9] As a teenager in Madras, Chandrasekhar had decided to pursue a research career in mathematics or physics, so in addition to his regular academic studies at Presidency College in Madras, he dove into an intensive self-directed study of atomic physics and statistical physics.

A 1928 visit to Madras by the eminent German physicist Arnold Sommerfeld (1868–1951) brought the recently derived Fermi-Dirac statistics to the young university student's attention. Browsing through some recently arrived journals in the university library, Chandrasekhar subsequently came across Fowler's paper on the dense matter in white dwarfs, and he began a correspondence with Fowler. Upon graduating with top honors in 1930, he wrote to Fowler asking for the opportunity to study physics with him at Cambridge University, and Fowler agreed.

[9] Chandrasekhar's early life is discussed by Holberg (2007), pp. 125 ff.

On July 31, 1930, the 19-year-old Chandrasekhar boarded a steamer from Bombay to Venice *en route* to England. With time on his hands during the long sea voyage, he began to ponder how to generalize Fowler's treatment of the matter in a white dwarf to include the effects of special relativity. The extreme relativistic limit, corresponding to electron energies much larger than mc^2, where the symbol m represents the electron mass, was an easy case to treat, but the result was surprising. In this limit, Chandrasekhar found that the pressure at zero temperature corresponded to an unstable polytrope, with a unique mass but with indeterminate radius. Above this mass, gravitational forces overwhelm the pressure support, and no stable white dwarf can exist. Now called the "Chandrasekhar limit," this mass is about 1.4 times the mass of the Sun. The existence of this limit has had a profound impact upon theories of stellar evolution.

However, the British academicans did not know what to make of the young Indian student's surprising result. Chandrasekhar had written a paper about his work, which he showed to Fowler shortly after arrival at Trinity College in Cambridge. Fowler sent it to astrophysicist Edward A. Milne (1896–1950) at Oxford, but Milne apparently did not grasp the significance of the limiting mass, either. In the end, Chandrasekhar sent the paper to the *Astrophysical Journal* in the United States, where it was finally published in 1931. The examiners for Chandrasekhar's Ph.D. thesis defense in 1933 were Fowler and Sir Arthur Stanley Eddington, by then the dominant astrophysicist of the early twentieth century.

During a subsequent postdoctoral fellowship at Cambridge, Chandrasekhar continued his work on the pressure-density relation for a fully degenerate electron gas, and in 1935 he gave a paper on his work at the monthly meeting of the Royal Astronomical Society.[10] After Chandrasekhar had finished his presentation, Eddington took the podium to speak on the topic of "Relativistic Degeneracy." While acknowledging that Chandrasekhar's mathematical derivation was without flaw, Eddington nevertheless

[10]The interactions between Chandrasekhar and Eddington are discussed by Holberg (2007), pp. 130 ff.

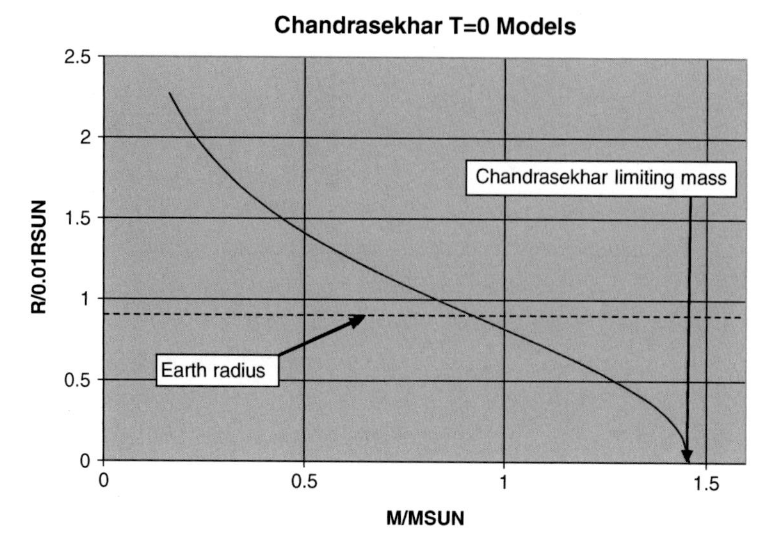

FIGURE 4.3 The theoretical mass-radius relation for fully degenerate white dwarfs, according to Chandrasekhar. Data for Chandrasekhar models from Hamada and Salpeter (1961)

argued[11] that "... there is no such thing as relativistic degeneracy!" Eddington questioned Chandrasekhar's method of combining quantum statistics with a relativistic energy-momentum expression (which Chandrasekhar and others subsequently vindicated), essentially ridiculing Chandrasekhar's results as irrelevant. One can only imagine what a devastating experience this must have been for the 24-year-old Indian astrophysicist! Chandrasekhar was vindicated in 1983 with the award of a Nobel Prize for this work, but Eddington, who remained unconvinced for the rest of his life, did not live to see it.

In his postdoctoral work, Chandrasekhar had extended his earlier calculations of the properties of degenerate matter, allowing the Fermi momentum of the electrons to extend continuously from the extreme non-relativistic limit to the extreme relativistic case. He was able to derive exact mathematical formulae to express the relation between the pressure and density, although the inclusion of this equation of state required numerical computations to obtain the resulting structures of the stars.

[11] Holberg (2007), p. 133.

In 1937, after settling into a position at the University of Chicago's Yerkes Observatory, Chandrasekhar included a detailed exposition of his white dwarf work in his classic 1939 monograph.[12] This work provided the first complete discussion of the equation of state at all densities and of the resulting structures of the zero temperature models. It established the existence of a mass-radius relation for fully degenerate stars that varies smoothly from the non-relativistic results found earlier by Fowler through smaller and smaller radii as the mass of the white dwarf increases, finally vanishing altogether at the Chandrasekhar limit (see Figure. 4.3 and Appendix C in this book).

[12]Chandrasekhar (1939), pp. 412 ff. A table of Chandrasekhar's zero-temperature white dwarf models for a range of masses is given on p. 427.

5. Star Power

By the end of the second decade of the twentieth century, the major problem remaining in the early study of stellar structure concerned the source of energy needed to supply the prodigious power output of the stars. Release of chemical energy, such as the burning of a lump of coal, was easily shown to be inadequate. The heat produced by complete combustion of a lump of coal would release only $2\text{–}3 \times 10^{11}$ erg g^{-1}, sufficient to maintain the current luminosity of the Sun only for 60,000–80,000 years. This is much less than the geological age of Earth.

In 1854, in a popular lecture in Königsberg, Hermann von Helmholtz (1821–1894) considered the energy that could be released by contraction of the Sun. To provide sufficient energy to sustain the solar luminosity at its present level would require a contraction rate of only about 7×10^{-5} cm s^{-1}. Although this does not sound like much, it meant that the Sun would have had to contract by an amount equal to its entire radius in only about 30 million years[1], again far smaller than the geological age of Earth. Neither internal heat nor energy supplied by hypothetical meteoritic bombardment was any more successful; both resulted in similar ages of a few tens of millions of years. Consequently, in 1861 Lord Kelvin concluded[2] that "neither the meteoric theory of solar heat nor any other natural theory can account for the solar radiation continuing at anything like the present rate for many hundreds of millions of years."

In 1905, a potential new source of energy was revealed by Einstein's famous equation $E = mc^2$. This implies that a gram of mass is equivalent to about 9×10^{20} erg of energy, more than a billion times larger than the amount that could be released by chemical

[1] Pannekoek (1989), p. 397.

[2] Chandrasekhar (1939), p. 484.

H.M. Van Horn, *Unlocking the Secrets of White Dwarf Stars*,
Astronomers' Universe, DOI 10.1007/978-3-319-09369-7_5

combustion. If some way could be found to convert a portion of the mass of the Sun into energy, the problem of the source of solar energy would be solved. Rutherford's 1911 introduction of the nuclear model for the atom, in which most of the mass resides in the small, positively charged nucleus, showed that it would somehow be necessary to tap the mass-energy of the atomic nucleus, and in the 1920s, Eddington and others speculated that nuclear energy sources might be the answer. At the time no one knew how this might be done, however.

Erwin Schrödinger's development of the wave theory of quantum mechanics in the mid-1920s opened a new possibility. Although—according to classical mechanics—two positively charged nuclei moving at speeds corresponding to the ten million Kelvin temperatures at the centers of Main Sequence stars would never be able to approach each other closely enough for a nuclear reaction to occur, quantum mechanics allowed a small possibility for this to happen. This was pointed out by Robert d'E. Atkinson (1898–1982) and Freidrich G. Houtermans (1903–1966) in 1929,[3] and a few years later particle accelerators showed experimentally that such nuclear reactions actually do occur.

Russian physicist George Gamow (1904–1968) calculated theoretically the penetration of charged particles through such nuclear potential barriers,[4] and the experimental results agreed with his calculations. He showed that, even at bombarding energies well below the maximum of the repulsive electrostatic potential barrier—proportional to the product of the positive electrical charges of two colliding nuclei—there is a finite but small probability of finding the two nuclei close enough together for a nuclear reaction to occur. For a given energy, the probability is greater when the product of the nuclear charges is small, and the probability increases rapidly as the relative energy of the colliding particles increases.

The next step in the elucidation of stellar energy sources was made by Hans Albrecht Bethe (1906–2005; see Figure 5.1). With a phenomenal memory and a prodigious work ethic, Bethe had developed a broad command of the field of physics, rapidly earning a reputation as one of the world's leading theoretical physicists. Born in

[3] Bethe (1991), p. 246.

[4] *Ibid.*

FIGURE 5.1 Hans A. Bethe in 1935. Photograph by Samuel Goudsmit, courtesy AIP Emilio Segré Visual Archives, Goudsmit Collection

Strasbourg, Germany, Bethe began his university studies in Frankfurt in 1924. His outstanding abilities were soon recognized, leading him to move to Munich to study under the eminent German physicist Arnold Sommerfeld. He received his doctorate in 1928, and after a postdoctoral interlude at the Cavendish Laboratory in England, he joined the faculty of Tübingen University in 1932.

In 1933, Bethe was part of the wave of anti-Semitic academic dismissals carried out in Nazi Germany. His mother was Jewish, though Bethe was raised as a Christian, but he never regarded himself as being very religious. The following year, Cornell University, in Ithaca, New York, offered Bethe an appointment as an assistant professor, which he accepted, arriving in 1935.

In 1937, Carl Friedrich von Weizsäcker (1912–2007) suggested as a possible source of nuclear energy for the Sun the simplest of all nuclear fusion reactions:

$$^{1}\mathrm{H} + ^{1}\mathrm{H} \rightarrow ^{2}\mathrm{H} + \mathrm{e}^{+} + \nu,$$

where the letters denote the chemical symbols for two different isotopes of the element hydrogen (H). The superscripts are the atomic masses of the nuclei; the symbol ^2H represents deuterium, or "heavy hydrogen," a nucleus consisting of a proton and neutron bound together by the strong nuclear force that was just then beginning to be understood. The remaining symbols represent positrons (e^+), which are positively charged electrons, and neutrinos (ν).

In 1938, Bethe and Charles L. Critchfield (1910–1994) published a theoretical calculation of the rate of this reaction,[5] which is much too slow to be observed in terrestrial laboratories. Immediately after it is formed, however, the deuteron is rapidly transformed into a ^3He nucleus by interacting with another proton, and then several additional reactions—collectively termed the "proton-proton chain"—complete the process, ending with the production of ^4He. Bethe and Critchfield found that the rate of energy production by these processes is indeed sufficient to supply the power radiated by the Sun, but the temperature-dependence of this thermonuclear reaction rate is not steep enough to account for energy production in more massive stars, which were known to radiate at a vastly greater rate than the Sun.

This was the situation in April 1938, when Gamow—by then a professor at George Washington University—brought together a small group of physicists and astrophysicists for a conference at the Department of Terrestrial Magnetism of the Carnegie Institution of Washington, in Washington, DC. Bethe was one of the participants. At this conference, the astrophysicists described for the physicists the current state of knowledge about the internal structure and constitution of the stars. Among other things, they showed that the central temperature and density of the Sun are approximately 14 million degrees Kelvin and 160 g cm^{-3}, respectively; that the Sun consists mostly of hydrogen, with some helium and a few percent of heavier elements; and that the energy sources for the more massive stars must be orders of magnitude larger than for the Sun.

Already fascinated with the problem of stellar energy, Bethe pondered on the train back to Ithaca what nuclear reactions might release energy at such prodigious rates. According to a popular

[5] *Ibid*, pp. 246–253.

book by Gamow,[6] when the first call for dinner on the train came, Bethe did not have the answer. Since he had a hearty appetite, however, he re-doubled his efforts, and by the time the porter came through the train to announce the final call for dinner, Bethe had the solution! When asked about this story decades later, after he had received the 1967 Nobel Prize for this work, Bethe grinned and replied, "The truth of the matter is that it took six weeks of hard work in the library at Cornell before I had the answer!"

The solution Bethe found[7] was the following sequence of nuclear reactions:

$$
\begin{aligned}
&{}^{12}\mathrm{C} + {}^{1}\mathrm{H} \rightarrow {}^{13}\mathrm{N} + \gamma \\
&{}^{13}\mathrm{N} \rightarrow {}^{13}\mathrm{C} + e^{+} + \nu \\
&{}^{13}\mathrm{C} + {}^{1}\mathrm{H} \rightarrow {}^{14}\mathrm{N} + \gamma \\
&{}^{14}\mathrm{N} + {}^{1}\mathrm{H} \rightarrow {}^{15}\mathrm{O} + \gamma \\
&{}^{15}\mathrm{O} \rightarrow {}^{15}\mathrm{N} + e^{+} + \nu \\
&{}^{15}\mathrm{N} + {}^{1}\mathrm{H} \rightarrow {}^{12}\mathrm{C} + {}^{4}\mathrm{He}.
\end{aligned}
$$

In these equations, the letter symbols are the chemical symbols for the elements carbon (C), hydrogen (H), nitrogen (N), and oxygen (O), and the superscripts are the atomic masses of the nuclei. In short, this sequence of reactions converts four protons (^{1}H) into the nucleus of one helium atom (^{4}He), in the process liberating energy in the form of gamma rays (γ), positrons (e^{+}), and neutrinos (ν). The gamma rays are quickly absorbed and converted into thermal energy, and the positrons rapidly annihilate with ambient electrons, producing additional gamma rays that are also absorbed. The ghostly neutrinos interact so weakly with matter that they escape almost completely unhindered from the solar interior.

These reactions in effect make use of carbon as a catalyst to combine four protons together to produce a ^{4}He nucleus (called an "alpha particle"). At the end of the sequence, a ^{12}C nucleus is regenerated, ready to start the sequence all over again. In addition to this set of reactions proposed by Bethe, another interlocking

[6] Gamow (1945).

[7] Bethe (1991), pp. 246–253.

cycle of catalytic reactions involving various isotopes of carbon, nitrogen, and oxygen produces energy and ^{4}He as well.

All of the reactions in these sequences—collectively termed the "CNO bi-cycle"—are "exoergic," meaning that they release energy, and the reactions proceed spontaneously in the directions indicated by the arrows. From Einstein's formula $E = mc^2$, the total energy released in the process is the difference between the sum of the masses of the four protons and the mass of the ^{4}He nucleus produced by either the proton-proton chain or the CNO bi-cycle sequences of nuclear reactions. Although this energy is only about 0.007 of the combined rest-mass energy of the four protons, it is still sufficient to power the Sun for a very long time.

If this fraction of, say, 10 % of the mass-energy of the Sun were to be liberated and radiated away at a rate corresponding to the current solar luminosity, for example, it would be sufficient to power the Sun for about 10 billion years. From radiogenic age measurements of the oldest terrestrial rocks, meteorites, and lunar samples, we now know that the age of the Solar System—including the Sun and planets—is about 4.5 billion years, so the nuclear energy source at the center of the Sun is indeed sufficient to power our own star for its entire lifetime.

What about other Main Sequence stars? From his investigations of stellar structure, Eddington had shown by quite general arguments[8] that the luminosity of a Main Sequence star increases very rapidly with its mass and that it depends very little on any other physical properties. Observational data confirmed this conclusion.

Now, with highly temperature-sensitive thermonuclear reactions "burning" hydrogen into helium deep in the central core of a star identified as the source of stellar luminosity, the underlying reason for this relationship became clear. The more massive stars have increasingly greater central temperatures, which in turn burn hydrogen much more rapidly, supplying very much higher luminosities. This is illustrated in Figure 5.2,[9] which shows the locus of points corresponding to the epoch where stars of different masses begin their hydrogen-burning Main-Sequence phase—the so-called "Zero Age Main Sequence," or ZAMS. For each star, this

[8] Pannekoek (1989), p. 465, Holberg (2007), pp. 109–112.

[9] Data from Schwarzschild (1958), p. 264, plotted by the author.

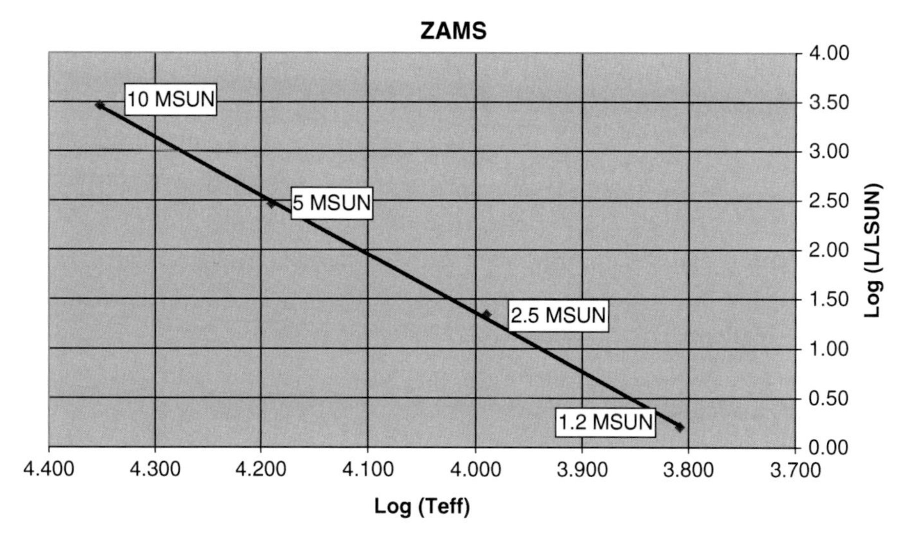

FIGURE 5.2 The Zero-Age Main Sequence in the Hertzsprung-Russell diagram. Luminosity increases upward from about the luminosity of the Sun to 10,000 times brighter. Effective temperature increases to the *left*, from about 5,000 K to about 25,000 K. Positions are shown for four models, with masses ranging from $1.2 M_{Sun}$ up to $10 M_{Sun}$, just at the beginning of hydrogen burning on the Main Sequence. Data from Schwarzschild (1958, p. 264)

figure gives the logarithm of the stellar luminosity at the beginning of hydrogen burning, in units of the luminosity of the Sun (log $[L/L_{Sun}]$) plotted against the logarithm of the stellar surface temperature (log $[T_{eff}]$), which astronomers call the "effective temperature." This plot is a theoretician's version of the famous Hertzsprung-Russell diagram.

In the original H-R diagram, the quantities plotted were observational ones—the absolute magnitude (instead of log L/L_{Sun}) and the spectral class or color (instead of log T_{eff}). The hotter O spectral class or blue color was traditionally plotted at the left of the horizontal axis, while the cooler M spectral class or red color was plotted to the right. This tradition explains why astronomers plot log (T_{eff}) *de*creasing to the right; they wanted the form of the plot to match the orientation of the original H-R diagram.

How long can hydrogen burning sustain a Main Sequence star? We saw earlier that it can sustain the Sun for many billions of years, but what about stars of other masses? At the low-mass extreme, an M-dwarf star, with about a tenth of the mass of the Sun, has a luminosity about one thousandth of the solar luminosity.

Even though it has only about a tenth as much hydrogen fuel as the Sun, its "hydrogen-burning" lifetime would thus be about 100 times longer than the Sun's because it radiates energy away so much more slowly. Thus, once such a low-mass star is formed, it consumes its hydrogen fuel very slowly, remaining essentially unchanged throughout the lifetime of the Universe.

Conversely, for an O star perhaps 40 times more massive than the Sun, the luminosity may be as much as 100,000 times the solar luminosity, or more. This enormous power output cannot be sustained by hydrogen-burning nuclear reactions longer than a few million years. The low-mass stars are powered mainly by the reactions of the proton-proton chain, while the massive, short-lived, high-luminosity stars are powered mainly by the CNO bi-cycle that Bethe had discovered.

This immediately leads to the natural question: What happens to a Main Sequence star when the hydrogen fuel in its core is exhausted?

Without a further nuclear energy source immediately available to maintain the core temperature, the H-exhausted stellar core begins to contract under gravity. This liberates gravitational energy while at the same time compressing and further heating the core. Immediately outside the He core—comprised of the "ashes" left behind by hydrogen burning—a hydrogen-burning shell continues to produce energy, and the two energy sources—gravitational energy released by the contracting core and nuclear energy released by the H-burning shell source—combine to produce an increasing power output that swells the outer layers of the star. That is, the star leaves the Main Sequence and begins to expand. Figure 5.3 shows this early post-Main-Sequence phase in the evolution of a 1.2 solar mass star with low abundances of heavy elements. As the surface of the star expands, the effective temperature declines, and the star becomes increasingly red.

Because the surface of the star expands so much, even though it has a lower effective temperature than it had on the Main Sequence, it reaches a very much higher luminosity. It is well on its way to becoming a "red giant." These high-luminosity red stars are the second-most populous group in the H-R diagram after the Main Sequence.

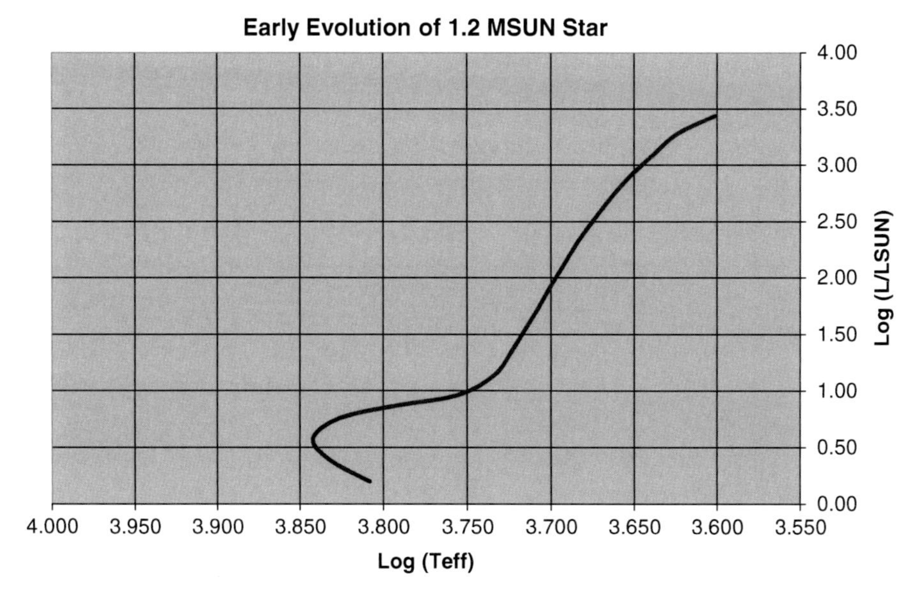

FIGURE 5.3 The early evolution of a 1.2 solar mass star with low abundances of heavy elements. Luminosity increases upward from about the luminosity of the Sun to 10,000 times brighter. Effective temperature increases to the *left*, from about 3,500 to 10,000 K. As hydrogen is exhausted in the core of the star, it moves to higher luminosities in the H-R diagram. The extended rise from log $(L/L_{Sun}) = 1.00$ to log $(L/L_{Sun}) = 3.50$ represents the star's ascent along the red giant branch (RGB). Data from Schwarzschild (1958), p. 264

In the mid-to-late 1950s, when the calculations in Figure 5.3 were carried out, all of them had to be done laboriously by hand.[10] A single numerical stellar model might consist of perhaps 100 shells, each of which had to have all of the physical properties calculated. And an evolutionary track like that shown in the figure might require scores of numerical models. The vast amount of mind-numbing effort required meant that few such evolutionary tracks had been calculated by the end of the 1950s. The advent of digital computers radically changed all that, and we'll return to this in a later chapter.

For now, however, let us continue to follow a star as it ascends the "red giant branch," or RGB, in the H-R diagram. What

[10] Schwarzschild (1958), pp. 199ff.

happens next? The H-burning shell continues to process H into He, adding to the mass of the H-exhausted core, and gravitational contraction of the core continues to heat it to higher and higher temperatures. Since the core is now almost pure ^{4}He—apart from a smattering of heavier elements—the obvious idea is to consider this reaction:

$$^{4}\mathrm{He} + {}^{4}\mathrm{He} \leftrightarrow {}^{8}\mathrm{Be}.$$

Unfortunately, the ground state of the ^{8}Be nucleus is 92 keV *higher* in energy[11] than the rest mass of the two alpha particles, making it unstable. For this reason, it spontaneously breaks up almost immediately into two alpha particles, as indicated by the double-ended arrow.

In 1952, Edwin E. Salpeter (1925–2008; see Figure 5.4) proposed a way around this problem. Another physicist with an interest in astronomical problems, he had been born in Austria. As a teenager, he had fled Nazi Germany with his family, emigrating to Australia and earning his bachelor's degree from Sydney University in 1944. After receiving a master's degree in 1945, he was awarded a scholarship to Birmingham University in England, where in 1948 he earned his Ph.D. under the direction of Sir Rudolf Peierls (1907–1995).

Salpeter went to Cornell in 1949 as a postdoc and soon joined Bethe on the Cornell faculty. What he pointed out in 1952 was that the short but finite lifetime of the ^{8}Be nucleus meant that the reaction would reach an equilibrium state in which a tiny concentration of ^{8}Be nuclei is present in the "sea" of ^{4}He nuclei. A second reaction could then take place:

$$^{4}\mathrm{He} + {}^{8}\mathrm{Be} \rightarrow {}^{12}\mathrm{C} + \gamma.$$

Since the ground-state energy of ^{12}C is 7.366 MeV *below* the combined energies of the ^{4}He and ^{8}Be nuclei, this second reaction does proceed spontaneously in the direction indicated by the

[11] An electron volt, abbreviated "eV," is a convenient measure of energy for use in discussing electronic levels within atoms. One million electron volts, or 1 MeV, which equals 1.602×10^{-6} erg, turns out to be a more convenient unit of energy to use in measuring nuclear or relativistic energy changes. One kilo electron volt, or 1 keV, equals 1,000 eV or one thousandth of an MeV.

FIGURE 5.4 Edwin E. Salpeter. Photograph by Russ Hamilton, Cornell University, Division of Rare and Manuscript Collections, Cornell University Library

arrow, releasing energy in the process. Because the two reactions combine three alpha particles into one ^{12}C nucleus, they are collectively termed the "3α" or "triple-alpha" reaction.[12] The 3α reaction is followed quickly by the further reaction

$$^{4}He + ^{12}C \rightarrow ^{16}O + \gamma,$$

so the products of He-burning end up being some mixture of carbon and oxygen. The central temperature necessary for a star to begin producing energy by means of these reactions is about an order of magnitude larger than the core temperatures in Main Sequence stars. Just such high temperatures are reached in the

[12] Bethe (1991), pp. 260–261.

core of a star when it reaches the tip of the red giant branch. Accordingly, we can identify the red-giant tip as the site of ignition of He-burning.

There is an interesting coda to this story.[13] The rate Salpeter originally calculated for the 3α reaction was not sufficiently fast to produce the observed relative abundances of ^{4}He, ^{12}C, and ^{16}O. Spectroscopic investigations by this time had shown not only that the stars and planets are composed of the same elements as those that appear on Earth, but also that there is a common pattern to the relative abundances of all the elements—the so-called "cosmic abundances." This led British astronomer Fred Hoyle (1915–2001)—later knighted to become Sir Fred—to propose that the ^{12}C nucleus must have an excited nuclear state at an energy 7.644 MeV above the ^{12}C ground state. If this were actually the case, then the probability of the second step in the 3α reaction would be greatly increased, producing relative abundances in agreement with measured cosmic element abundances of ^{4}He, ^{12}C, and ^{16}O.

The American nuclear astrophysicist William A. Fowler (1911–1995) and his colleagues at Caltech's Kellogg Radiation Laboratory subsequently carried out careful measurements of the internal states of the ^{12}C nucleus and found an excited state with almost precisely the properties Hoyle had predicted. For their insightful work on this key problem, Salpeter and Hoyle were jointly awarded the Crafoord Prize by the Royal Swedish Academy of Sciences in 1997.

What implications does all of this have for white dwarfs? First, calculations of the evolution of nuclear-powered stellar models had shown by the late 1950s that if a star is sufficiently massive to leave the Main Sequence—and thus in some as yet undetermined way eventually become a white dwarf—it will no longer consist primarily of hydrogen. Second, since all such models ignite He-burning, a white dwarf probably contains a substantial amount of carbon and oxygen, but maybe not much helium. We shall see later that both of these suppositions are correct, but first we need to look at the thermal properties of white dwarfs.

---

[13] *Ibid*, p. 261.

6. Still Pretty Hot for a Fading Old Star!

Chandrasekhar's models assumed white dwarfs to be fully degenerate—that is, effectively at zero temperature. Of course, no real star has temperature of zero. The reality that a white dwarf emits observable radiation means that the surface temperature not only is non-zero but in fact is rather high.

From the Stefan-Boltzmann law, a white dwarf with a radius about a hundredth of the solar radius and a luminosity about a hundredth of the solar luminosity has a surface temperature about three times greater than that of the Sun, or roughly 18,000 K. And if the surface temperature is that high, the internal temperature must be higher still. The relation between temperature gradients and luminosities obtained with conventional radiative opacities briefly led astrophysicists in the early twentieth century to conclude that the internal temperatures might actually be as high as a billion degrees Kelvin! Even at densities as large as those in typical white dwarfs, electrons are not degenerate at such high temperatures. This picture clearly was not self-consistent!

On the other hand, Chandrasekhar's fully degenerate models do produce a mass-radius relation that agrees with observations, suggesting that the internal thermal energies of white dwarfs really are small compared with the energies of degenerate electrons. The flaw in the reasoning that had led to the inference of incredibly high white-dwarf internal temperatures was the fact that these calculations had failed to take account of the very high thermal (and electrical) conductivity of degenerate matter.

Robert E. Marshak (1916–1992; see Figure 6.1) first computed the conductivity of degenerate matter for conditions appropriate

H.M. Van Horn, *Unlocking the Secrets of White Dwarf Stars*,
Astronomers' Universe, DOI 10.1007/978-3-319-09369-7_6

Figure 6.1 Robert E. Marshak, *circa* 1950. Credit: AIP Emilio Segré Visual Archives, Physics Today Collection

to the deep interior of a white dwarf.[1] He was born and grew up in the South Bronx, his Jewish parents having emigrated from Czarist Russia to escape oppression.[2]

In 1932, Marshak enrolled in the City College of New York and, after completing his undergraduate degree, began his graduate study in physics at Cornell University, working under Hans Bethe.

Knowing that Chandrasekhar had shown white dwarfs to be degenerate stars but that the internal temperatures were still a mystery, Bethe assigned Marshak the thesis problem of calculating the thermal conductivity of white dwarf matter. Bethe had already made significant contributions to solid-state physics, and he recognized the similarity between the electron degeneracy in white dwarfs and that in ordinary terrestrial metals. In both cases, the electrons are so highly degenerate that the Pauli Exclusion

[1] Holberg (2007), p. 149.

[2] Marshak (1982), p. xi.

Principle inhibits scattering of the electrons by the ions, because an electron can only be scattered into an unoccupied final state. Electrons can therefore travel long distances between scatterings under these conditions, and for this reason electron conduction is highly efficient in transporting energy.

Following this solid-state-physics approach, Marshak soon completed a calculation of the thermal conductivity of white dwarf matter. He was then able to relate the heat flux passing through an imaginary spherical shell in a white dwarf to the local temperature gradient.

To put his result in a form more familiar to astronomers, Marshak took advantage of the fact that the temperature-gradient equation used in stellar structure calculations has exactly the same form as the heat flow equation, except that the factor of proportionality in the former is written in terms of the opacity of stellar matter. Marshak used this equivalence to translate the thermal conductivity he had obtained into a form he called the "conductive opacity," which could be used directly in stellar structure calculations. And because thermal conduction by degenerate electrons and radiative transfer by photon diffusion provide independent, parallel pathways for energy transport, the larger the conductivity (or, equivalently, the lower the conductive opacity), the more efficiently energy is transported through the stellar matter.

To find the temperature distribution in a white dwarf, Marshak calculated an approximate solution to the equations of stellar structure, using a previously computed expression for the radiative opacity and his own result for the conductive opacity.[3] Employing a model for Sirius B, he showed that the conductive opacity remains many factors of ten smaller than the radiative opacity throughout the strongly degenerate interior of a white dwarf, making electron conduction overwhelmingly the dominant form of energy transport throughout the white dwarf core. The conductive opacity exceeds the radiative opacity only far out near the stellar surface, where the declining density causes the Fermi energy to fall below the thermal energy at a concentric spherical

[3] Marshak, Robert E. 1940, *Astrophys. J.*, **92**, 321, "The Internal Temperature of White Dwarf Stars."

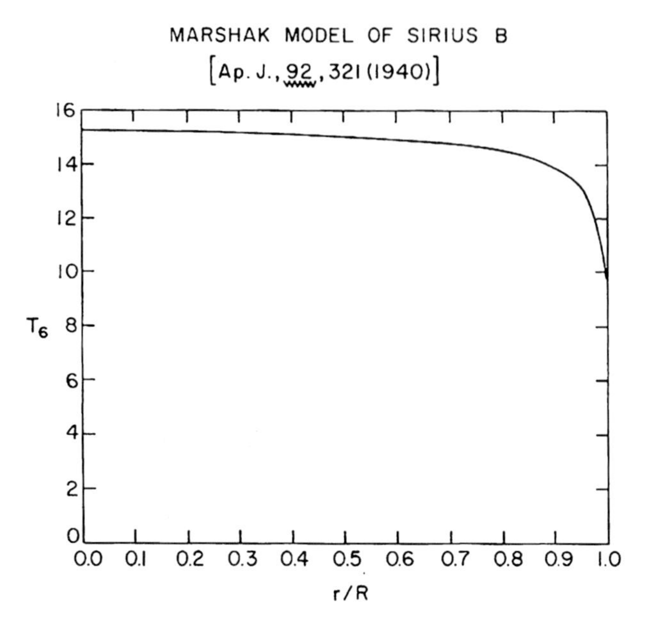

Figure 6.2 Temperature distribution in the white dwarf Sirius B, according to 1940 calculations by Marshak. The vertical axis gives the temperature in units of 10^6 K, while the horizontal axis is the fraction of the surface radius of the white dwarf. The central temperature is slightly more than 15 million degrees Kelvin, and the temperature does not drop below about 14 million Kelvin until we reach about 90 % of the distance to the stellar surface. From Van Horn (1971). © IAU. Reproduced with permission

shell termed the "degeneracy boundary." The white dwarf core—like a metal under normal laboratory conditions—is an extremely good conductor of heat. Marshak thus found that the vast bulk of the degenerate interior of a white dwarf has a nearly uniform temperature of about 10 million degrees Kelvin (10^7 K) (see Figure 6.2).[4] That is, to a good approximation, the interior of a white dwarf star is almost isothermal. The thin, non-degenerate surface layers, spanning only a few percent of the radius of the star, act as an efficient insulating blanket surrounding this hot "metallic" core, and most of the temperature drop between the center and the surface occurs across this thin layer.

[4]Data from Van Horn, H. M. (1971), in Luyten (1971), p. 97, "Cooling of White Dwarfs."

Based on Marshak's results, it is easy to justify the use of zero-temperature models for the mechanical structure of a white dwarf: For a central density of about a million g cm^{-3}, the Fermi energy of the degenerate electrons is about 1 Mev. In contrast, a temperature of 10 million Kelvin corresponds to an energy of only about a thousand electron volts, or 1 keV, a thousand times smaller than the energies of the degenerate electrons.

After receiving his doctorate from Cornell in 1939, Marshak joined the physics faculty at the University of Rochester in Rochester, New York. But he left almost immediately to join Bethe and other eminent scientists working on the Manhattan Project at the top-secret Los Alamos National Laboratory in New Mexico in the race to develop the atomic bomb and end World War II. After the war, Marshak returned to Rochester, where he soon became chairman of the physics department, focusing his considerable energies in the then-new field of elementary particle physics. He established the Rochester Conferences to promote international scientific cooperation in this subject during the Cold War, and he made major contributions to meson physics and the theory of the weak interactions among elementary particles. In 1970 he left Rochester to become president of his alma mater, CCNY, retiring in 1979 to Virginia Polytechnic and State University, in Blacksburg, Virginia. He died unexpectedly in 1992, drowning while swimming in Cancun, Mexico.[5]

Although the core temperatures of the white dwarfs had been established by Marshak's calculations, the nature of a white dwarf's energy source remained a mystery until the mid-twentieth century. As civilization gradually began to recover from the devastations of World War II and scientists returned to their peacetime pursuits, the puzzle of white dwarfs again began to occupy the thoughts of astronomers. White dwarf core temperatures are certainly high enough to sustain thermonuclear burning of hydrogen, if any hydrogen were to exist in the deep interior. But the densities inside these degenerate stars are so high that such burning would have generated many thousands of times as much energy as white dwarfs were observed to radiate.

---

[5] Melissinos and Van Horn (1993), pp. 1–3.

Furthermore, such burning would be dramatically unstable, as shown by T. D. Lee (b. 1926) in 1950.[6] If the energy-generation rate is too low, a degenerate star cannot contract and heat, and the energy production simply dies. Conversely, any excess energy production cannot expand the star to shut off the nuclear reactions, so that a thermal runaway results. Such explosive thermonuclear outbursts are indeed observed as nova explosions[7] in white dwarfs that have accreted sufficient hydrogen-rich fuel from a companion star, but these events cannot be the source of energy for isolated white dwarfs. And since the central temperatures of white dwarfs are too low for helium burning or any later stages of thermonuclear burning, it was clear that the source of white dwarf luminosity could not be due to nuclear reactions.

An important corollary to these results was the conclusion that the amount of hydrogen in a white dwarf must be nearly zero, even in the non-degenerate envelope, despite the fact that hydrogen lines are clearly seen in the spectra of a majority of these stars. If as much as one ten-thousandth of a solar mass of hydrogen were present in an isolated white dwarf, hydrogen-burning nuclear reactions would proceed explosively. That this does not occur provides yet another argument for the advanced evolutionary status of white dwarfs.

Another possibility was that energy might be released by the gravitational contraction of a white dwarf. However, if it truly were fully degenerate, the star would be supported against gravity by the pressure of the degenerate electrons, even at zero temperature, and any residual contraction would be negligibly small.

That left only the residual heat contained in the degenerate core of the white dwarf as the source of energy. Although this had long since been dismissed as the power source for the Sun, white dwarfs radiate their energy much more slowly than the Sun. Would the internal heat perhaps suffice for white dwarfs? In the early 1950s, British astrophysicist Leon Mestel (Figure 6.3) set about finding out. Born in Melbourne, Australia, in 1927, he moved to

---

[6] Lee, T. D. 1950, *Astrophys. J.*, **111**, 625, "Hydrogen Content and Energy-Productive Mechanism of White Dwarfs."

[7] A supernova explosion, to which we shall return later, is vastly more energetic than a nova outburst and involves the incineration and disruption of the entire star.

Figure 6.3 Leon Mestel in 1976. Credit: AIP Emilio Segré Visual Archives, John Irwin Slide Collection

London at age three with his family. In 1948 he received his bachelor's degree from Trinity College of Cambridge University and earned his doctorate 4 years later.

In 1952, Mestel proposed a simple, physical model[8]—in which the residual heat content of the degenerate core gradually leaked away through the thin, non-degenerate surface layers—to account for the luminosities and ages of white dwarfs. Although far more detailed calculations of white dwarf evolution have been done since Mestel's seminal work, his model contains the basic concepts, and a review of it provides a good introduction to current ideas about white dwarf cooling.

Since the non-degenerate envelope of a white dwarf is very thin, Mestel neglected its mass in comparison with the total mass M of the star. He also neglected energy production or loss within

[8] Mestel, L. 1952, *Mon. Notices Roy. Astron. Soc.*, **112**, 583, "On the Theory f White Dwarf Stars. I. The Energy Sources of White Dwarfs"; summarized by Van Horn 1971, *op. cit.*

these layers, taking the luminosity L to be constant throughout the envelope. Because these layers are non-degenerate, the properties of matter are those of an ideal gas. Using a conventional expression for the radiative opacity, Mestel was then able to solve the two remaining stellar structure equations to obtain the distribution of pressure as a function of temperature throughout these surface layers. The result depends upon the ratio of the luminosity L of the white dwarf to its total mass M.

Mestel next determined the relationship between the pressure and temperature at the degeneracy boundary. Because the Fermi energy of the degenerate electrons depends on the pressure, and because the degeneracy boundary is by definition the spherical shell within the white dwarf at which the Fermi energy equals the thermal energy, there is a unique relationship between the temperature and pressure at the degeneracy boundary.

Combining the pressure-temperature relation for the degeneracy boundary with that for the non-degenerate envelope produced a simple power-law relation between the luminosity of the white dwarf and the temperature at the boundary of the degenerate core. Mestel also realized that—because the degenerate electrons conduct heat so efficiently that the core of the white dwarf is nearly isothermal—the core temperature could be taken to be approximately equal to the temperature at the degeneracy boundary. Assuming that the only source of energy in a white dwarf is its heat content, which for degenerate matter is just that of the ions, he obtained an equation governing the time rate of change of the heat content of the white dwarf.

Mestel solved this equation using the luminosity-core temperature relation to obtain the time required for the white dwarf to cool to a given core temperature. For a pure ^{12}C white dwarf, for example, this cooling time turns out to be about 130 million years for a white dwarf to reach a core temperature of 10 million Kelvin. This does indeed provide the correct order of magnitude for white dwarf cooling times. A white dwarf cools increasingly slowly as the core temperature falls so that, for example, it takes more than 1.3 billion years to cool to a core temperature of about 4 million Kelvin.

Because there is a direct relationship between the core temperature and the luminosity of a white dwarf, Mestel was also able to calculate the cooling time to a given luminosity. For a one-solar-mass white dwarf like Sirius B, for instance, the cooling time to a luminosity of about 10^{-2} L_{Sun} is about 30 million years. And it takes 140 million years to fade to 10^{-3} L_{Sun} and more than 700 million years to cool to 10^{-4} L_{Sun}. The cooling time also depends somewhat on the mass of the white dwarf, with lower-mass stars cooling somewhat faster because of their larger radii: As one example, a white dwarf with a mass of 0.6 M_{Sun} fades to 10^{-3} L_{Sun} in less than 100 million years.

As these results show, it takes longer and longer for a white dwarf to cool to lower and lower luminosities. In other words, the rate of cooling decreases steadily the fainter the star becomes. This has an important observational consequence. If we plot the numbers of white dwarfs in any given interval of absolute magnitude (or equivalently, in any given interval in the logarithm of the luminosity), the number of white dwarfs becomes larger at lower luminosities. We shall return to this point and its consequences again in Chap. 21.

Following this seminal early work on white dwarfs, Mestel's interests turned to star formation and more general topics in stellar structure,[9] particularly issues concerning stellar magnetic fields and astrophysical magnetohydrodynamics. He made another important contribution to our understanding of the thermal properties of the white dwarfs in the late 1960s, to which we shall return in Chap. 11. He retired in 1992 as professor emeritus at the University of Sussex, having garnered a number of prestigious awards for his work.

[9] Mestel's career is summarized in http://en.wikipedia.org/wiki/Leon_Mestel, accessed April 23, 2012.

7. Stalking the Wild White Dwarf

During the first half of the twentieth century, while theorists were working to understand the structure and thermal properties of white dwarfs, observers began to search for more of these peculiar stars. Were they rare or commonplace? What were their observable properties?

Willem J. Luyten (1899–1994; see Figure 7.1) was the early leader in this effort. Born in the Dutch East Indies (now Indonesia) to a Dutch father and French mother,[1] he moved to the Netherlands as a teenager with his parents in 1912. After earning his Bachelor's degree from the University of Amsterdam in 1918, he became Ejnar Hertzsprung's first doctoral student at Leiden University and received his Ph.D. 3 years later.

Luyten's first postdoctoral appointment was at the Lick Observatory in California. Established on Mount Hamilton near San Francisco just 33 years previously,[2] this observatory then possessed the second-largest refracting telescope in the world. An 1876 bequest from wealthy Californian James Lick had provided funds to construct the facility. Final figuring of the high-quality, 36-in. glass blanks for the lenses was done by the highly respected optical firm of Alvan Clark & Sons, the same Alvan Clark and his son Alvan Clark, Jr., who were the first people in the world to have seen white dwarf Sirius B.

The new observing facility became the world's first mountaintop observatory. Construction began in 1880, with the telescope lenses delivered to the mountaintop by horse and wagon just

[1] A biographical sketch of Willem J. Luyten is given in "The Bruce Medalists," http://www.phys-astro.sonoma.edu/brucemedalists/luyten/index.html, accessed 26 April 2012.

[2] The history of Lick Observatory is summarized by Misch, A., and Stone, R. 1998, "Building the Observatory, http://collections.ucolick.org/archives_on_line/bldg_the_obs.html, accessed 22 May 2012.

H.M. Van Horn, *Unlocking the Secrets of White Dwarf Stars*,
Astronomers' Universe, DOI 10.1007/978-3-319-09369-7_7

FIGURE 7.1 Willem J. Luyten. Image provided by the late W. J. Luyten and made available courtesy Joseph. S. Tenn

before Christmas in 1886, and the telescope saw "first light" on a bitterly cold night in January 1888. The "Great Refractor" was the largest in the world until construction of Yerkes Observatory's 40-in. eclipsed it a decade later.

At the time of Luyten's appointment at Lick in 1921, only three stars were known to possess stellar masses crammed into planetary dimensions: Sirius B, 40 Eridani B, and van Maanen 2. Luyten coined the name "white dwarfs" to describe them, listing them in a catalog of stars with large "proper motions" that he published in 1923.

These three stars shared two key observational properties: they were among the faintest then known—thus requiring for their investigation the use of the largest telescopes, which have the greatest light-gathering power—and they were much hotter than ordinary stars with similar luminosities, giving them much bluer colors than the faint red stars at the lower end of the Main Sequence. To Luyten, this suggested an obvious search strategy.

Because white dwarfs are intrinsically faint, they must be much nearer than ordinary stars with similar apparent magnitudes. Indeed, the first three stars to be recognized as white dwarfs are all located within 5 pc of the Sun. Because they are closer, the natural motions of white dwarfs across the sky (termed their "proper motions") must be larger than those of ordinary stars with similar apparent magnitudes. Thus, a search for stars with large proper motions was likely to turn up significant numbers of white dwarf candidates, and this was the first strategy Luyten adopted. It became his life's work to determine the proper motions of hundreds of thousands of stars.

To determine proper motions, Luyten made painstaking comparisons of the positions of stars in the sky at two different epochs. The basic idea was simple. First, take a photograph of a given patch of sky at some point in time. Then return sometime later and take another photograph of exactly the same region. Compare the apparent positions of the stars in the two images, and measure the displacement between the two images of the same star. The more distant stars exhibit smaller proper motions than the nearer ones, so they can be used as a framework against which to measure the motions of the nearer stars.

To expedite these measurements, Luyten employed a machine called a "blink comparator," inventing substantial improvements to an existing device that greatly facilitated the work. The improved machine allowed him to switch rapidly back and forth between two photographic plates, each containing an image of the same identical patch of sky. Once the images of the distant stars had been aligned between the two photographs, the process of switching rapidly between them caused the images of the stars with larger proper motions to appear to jump back and forth. Using this device, Luyten set about measuring the proper motions for as many nearby stars as possible.

At the end of his 2-year appointment at Lick, Luyten accepted an invitation to move to a position at the Harvard College Observatory.[3] Established in 1839 as part of the U.S.'s oldest university, this observatory had already played a leading role in the development of stellar astrophysics, including the creation of a system of spectral classification for stars. Only a year into his new appointment,

[3] Luyten's move to the Harvard College Observatory and his loss of an eye in a tennis accident are noted in http://www.nndb.com/people/232/000170719/, accessed 26 April 2012.

Luyten lost the sight in one eye in a tennis accident; he was only 25 years old at the time. Despite the resulting handicap, Luyten remarkably was able to continue his visually demanding proper-motion studies, which required precise determinations of minute position changes in the sky.

In 1929, Luyten initiated the Bruce Proper Motion Survey, using photographic plates taken at Harvard's Bruce telescope in South America. The Harvard College Observatory had recognized early in its history the importance of observations from the Southern Hemisphere, since some celestial objects—such as the center of our own Milky Way Galaxy—can never be seen from the north. For this reason, just 6 years after its founding, the Harvard College Observatory had installed the Bruce photographic telescope at Arequippa, Peru.

Luyten's appointment at Harvard gave him access to the Bruce plates, which provided the first-epoch images for his proper motion survey. One of the first and most extensive of such efforts, it revealed hundreds of southern hemisphere white dwarf candidates. Luyten's Bruce Proper Motion (BPM) catalog, published in 1963, includes all of the Southern Hemisphere, albeit to brighter limiting apparent magnitudes than subsequent proper-motion catalogs. Two years later, Luyten and his wife of about a year moved to the University of Minnesota, where he spent the remainder of his long and productive career.

Luyten's proper-motion surveys ultimately spanned seven decades of work. His Luyten Half Second (LHS) catalog contains about 4,000 stars near the Sun with proper motions of one-half second of arc per year or more. Though most are not white dwarfs, an appreciable number of LHS stars are. Luyten followed this work with his Luyten Two Tenths (LTT) survey, which contains more than 10,000 stars having proper motions in excess of 0.2 arcsec year^{-1}. Final compilations with measured proper motions exceeding 0.5 and 0.2 arcsec year^{-1}, respectively, were published in 1979 as the (updated) Luyten Half Second Catalog (LHS) and the New Luyten Two Tenths catalog (NLTT). These catalogs included all proper-motion stars known at the time, ordered by right ascension and identified by serial LHS and NLTT numbers.

Luyten's remarkable decades-long effort culminated in the Luyten Palomar (LP) survey in the 1960s and 1970s. For the first-epoch measurements for this survey, he employed the red-sensitive photographic plates obtained with the Palomar Observatory's 48-in. Schmidt telescope.

In 1929, Estonian optician Bernhard Schmidt (1879–1935) had invented a telescope of new design that proved ideal for photographing large regions of the sky. The main mirror in a Schmidt telescope is spherically curved, rather than having the parabolic shape to which most reflecting telescopes are figured. Schmidt's genius consisted of providing a glass corrector plate (lens) at the front of the telescope to compensate for the spherical aberration that would otherwise be introduced by the mirror. Schmidt's unique design resulted in a telescope with a much larger field of view than a conventional reflecting telescope, and this made it a superb instrument for identifying targets for follow-up observations by larger telescopes with narrower fields of view. Completed in 1948, the 48-in. Schmidt telescope began the first Palomar Observatory Sky Survey (POSS1) during the following year. This survey eventually mapped the entire sky in the northern hemisphere. Luyten obtained an additional series of red plates during the early 1960s that served as the second-epoch plates for the LP survey.

By the end of his long career, Luyten had screened more than 100 million stars[4] using his blink microscope, finding many thousands of white dwarf candidates in the process. He ultimately persuaded the Control Data Corporation—a computer manufacturer—to construct an automated machine to scan and measure images on photographic plates,[5] rather than having to examine each stellar image by eye, and with this machine he determined the proper motions for some 400,000 stars.

A few years after he retired in 1967, Luyten submitted an application for a small grant to the National Science Foundation, seeking financial support for a conference he was organizing on white dwarfs. This application produced one of his favorite stories.[6] Somehow, the grant proposal prompted an admonitory letter

[4] Luyten (1971), p. 4.

[5] Saxon, Wolfgang 1994, "Willem J. Luyten, 95, Expert on Stellar Motions," http://www.nytimes.com/1994/11/26/obituaries/willem-j-luyten-expert-on-stellarr-motion ... " accessed 26 April 2012.

[6] Holberg (2007), p. 119.

from the Surgeon General advising him "that human subjects could not be used for experimentation and that Federal money could not be used for race discrimination." Evidently, someone in the bureaucracy had failed to understand that Luyten's proposal concerned small, hot stars and not Snow White's companions!

Luyten was a gruff man who never shied away from confrontations with other astronomers over perceived inequities in the treatment of his work, priorities in discovery, or inaccuracies. He relished his role as a self-proclaimed "astronomical curmudgeon." Because much of his work was focused on the discovery of white dwarfs, he was sometimes called a "stellar mortician." He died of heart failure in 1994 at age 95.

Henry Giclas (1910–2007) and his colleagues at Lowell Observatory initiated another important proper-motion survey in 1957.[7] Their work employed the specially designed 13-in. Cooke refractor used in the 1930s by Clyde W. Tombaugh (1906–1997) to discover Pluto. Tombaugh's photographic plate set, covering about three-fourths of the sky at declinations north of –40°, served as the first epoch for this proper-motion survey, with the second-epoch plates obtained in the 1950s and 1960s. The northern and (limited) southern hemisphere catalogs of the "G-numbered stars" were published in the 1970s. The designation system the Lowell astronomers employed meant that, for example, the star G 29–38 was the 38th proper-motion star measured in plate field number 29. These lists of stars with proper motions larger than 0.26 arcsec year^{-1} contributed hundreds of white dwarfs to the currently known sample. Additional "GD" lists of blue stars showing smaller but detectable proper motions included eleven stars in the Hyades star cluster as well as many hot white dwarfs.

A second strategy employed by astronomers to identify white dwarf candidates capitalized on the fact that the colors of white dwarfs are bluer than those of Main Sequence stars. Searches for such faint blue stars began with the work of Milton Humason (1891–1972) and Fritz Zwicky (1898–1974) at the Palomar Observatory in 1947. Their idea was simply to compare images on pairs of photographic plates, one sensitive to blue light and the other to

[7] Giclas, H. L. in Luyten (1971), p. 24, "The Identification of White Dwarf Suspects in the Lowell Proper Motion Stars."

red. Of the list of 48 faint blue stars they identified in this way, the 15 stars in the Hyades cluster proved to be mostly white dwarfs, while only two of the remaining 33 were.

Luyten followed up Humason's and Zwicky's work using plates taken at several other observatories, including the Tonanzintla Observatory in Puebla, Mexico. Construction at Tonanzintla had begun in 1941 in collaboration with the Harvard College Observatory.[8] Sited only 19° north of the equator, Tonanzintla provided observational access to much of the rich southern sky. From 1950 to 1975, Guillermo Haro (1913–1988) served as director of the Tonanzintla Observatory, and Luyten began a productive collaboration with him to search for faint blue stars. By the time Luyten retired, he had published more than 20,000 white dwarf suspects, 8,700 of them jointly with Haro.

Indeed, Luyten's candidates with bluer colors—as well as those turned up in later surveys at the Lowell Observatory—have had a high "yield" in white dwarfs. Luyten published several lists of stellar objects with blue colors, without regard to their astrometric properties. The Luyten Blue (LB) catalog is actually a 50-volume series, spanning the years 1955–1969. Generally entitled "A Search for Faint Blue Stars," these are mostly University of Minnesota Observatory publications. Luyten also reported proper-motion measurements and provided some identification charts in these publications.

From studies such as these proper-motion or color surveys, thousands of white-dwarf candidates had been identified by the middle of the twentieth century. However, new developments in technology were already underway that would completely transform the way astronomers would search for white dwarfs. This began with the application of electronic technology to astronomy, starting early in the twentieth century. In 1907, American astronomer Joel Stebbins (1878–1966) used a photoconductive device[9] to produce some impressively accurate photometric measurements.

The devices that became the major focus of astronomical effort for much of the twentieth century, however, made use of the

[8] The origin of the Tonanzintla Observatory is described by Silvia Torres-Peimbert in DeVorkin (1999), p. 74, "A Century of Astronomy in Mexico: Collaboration with American Astronomers."

[9] Light shining on such a device changes its electrical conductivity.

photoelectric effect, discovered in 1887. In 1910, the discovery that the hydrides of sodium and potassium were sensitive photoemitters led to the construction of photoelectric cells, which were soon applied to astronomy. Although rapid improvements in the technology continued—including cooling to the temperature of dry ice, amplification of the very weak signals by increasingly sophisticated vacuum-tube amplifiers, and especially the development of photomultiplier tubes in the late 1930s—almost all stellar photometry continued to employ photographic methods up to the beginning of World War II.

After the war, the commercial availability of photomultipliers such as the RCA 931A and 1P21 led to a rapid expansion in the use of photoelectric photometry. This was strongly aided by Harold L. Johnson's (1921–1980) and William W. Morgan's (1906–1994) development of a standardized system of photoelectric photometry in the early 1950s. Both initially used three broadband filters, each roughly 1,000 Å wide, to obtain measurements of the brightnesses of stars in three colors: an ultraviolet *U* filter centered at approximately 3,600 Å, a blue *B* filter at 4,400 Å, and a "visual" *V* filter at 5,500 Å. The system was subsequently expanded by Gerald Kron (1913–2012) in the United States and Alan Cousins (1903–2001) in South Africa by adding a red *R* filter peaking at 7,000 Å, and a near-infrared *I* filter at 8,800 Å.

Measurements taken through the *V* filter approximated the earlier visual apparent magnitudes, while those taken with the *B* filter approximated photographic apparent magnitudes. In addition to enabling much faster and more accurate measurements of the apparent magnitudes of stars, the measurement of three broadband colors also provided a fast and efficient way to determine the shape of the stellar spectrum and the star's effective temperature. The individual photoelectric magnitudes provided logarithmic measures of the flux of light from a star in the range of the particular filter being used, while differences between two magnitudes—such as B–V—could be calibrated against the stellar temperature. Furthermore, because the broadband colors gathered photons from relatively wide swaths of the spectrum, the new system could be applied immediately to faint stars such as white dwarfs.

In the late 1970s, Richard Green made excellent use of these technological advances in his Ph.D. thesis research under professor Maarten Schmidt at Caltech, which culminated in the Palomar-Green (PG) survey.[10] Using prudently chosen photometrically calibrated colors and magnitudes and a careful calibration of the red color cutoff, his systematic survey targeted faint blue objects—such as white dwarfs or quasars—that have a strong excess of flux at ultraviolet wavelengths. This proved extraordinarily efficient in identifying very hot, pre-white dwarf stars. The PG survey found degenerate stars hotter than about 8,000 K (together with other faint, blue, apparently stellar objects), while most of the Main Sequence stars in the fields Green studied were excluded.

This survey, using the Palomar 18-in. Schmidt telescope, covered approximately one-quarter of the entire sky at galactic latitudes north of 30° and includes 1,874 blue stellar objects. Classification spectra were acquired for each object, and photometric colors and more precise spectra were obtained for hundreds of stars, including many white dwarfs. Each star is designated by an abbreviated position in the sky. For example, the famous pulsating variable star PG 1159-035 (see Chap. 17) is found at approximately right ascension 11 h 59 min and declination –3.5° (or –3°, 30 arc min).

Several other, more recent surveys were carried out in an effort to enlarge the number of white dwarf candidates in the southern hemisphere. The Montreal Cambridge Tololo (MCT) survey[11] used the Curtis Schmidt telescope at Cerro Tololo Interamerican Observatory (CTIO) to cover the south Galactic polar cap, a region not reached by the PG survey but using similar criteria. The Edinburgh-Cape (EC) Blue Object survey,[12] begun in 1985, was also designed to discover objects in the southern sky. Blue stellar objects were selected by automatic techniques from photographic

---

[10] Green (1977); see also Green, R. F., Schmidt, M., and Liebert, J. W. 1986, *Astrophys. J. Suppl.*, **61**, 305, "The Palomar-Green Catalog of Ultraviolet-Excess Objects."

[11] Demers, S. *et al.* 1986, *Astron. J.*, **92**, 878, "The Montreal-Cambridge Survey of Southern Subluminous Stars;" see also Lamontagne, R., *et al.*, *Astron. J.*, **119**, 241, "The Montreal-Cambridge-Tololo Survey of Southern Subluminous Blue Stars: The South Galactic Cap."

[12] Stobie *et al.* in Warner (1992), p. 87, "The Edinburgh-Cape Blue Object Survey."

plates taken with the UK Schmidt telescope in Australia. Several additional color surveys also resulted in the discovery of many more white dwarfs. The Hamburg-Schmidt (HS) Survey[13] used a telescope relocated to the Spanish-German Astronomical Center at Calar Alto (Spain), is a wide-angle, objective prism survey of an area of more than a third of the sky), while the Hamburg-ESO (HE) survey[14] used the Schmidt telescope at the European Southern Observatory (ESO) in Chile in a similar fashion to observe thousands of square degrees in the southern part of the sky.

Even as astronomy was being transformed by electronics, electronics was undergoing a further transformation from vacuum-tube technology to solid-state devices. In 1947, Walter Brattain (1902–1987), John Bardeen (1908–1991), and William Shockley (1910–1989) at Bell Laboratories invented the first semiconductor device, the transistor. In 1958, Jack Kilby (1923–2005) at Texas Instruments figured out how to create an entire electronic circuit on a single wafer, or "chip," of silicon. Robert Noyce (1927–1990) at Fairchild Semiconductor independently made the same invention about the same time—and the integrated circuit was born.[15] With the passage of time, individual circuit elements became smaller, and the complexity of circuitry that could be installed on a chip became greater.

In 1969, Bell Labs scientists George Smith (b. 1930) and Willard Boyle (1924–2011) invented the charge-coupled device (CCD),[16] which was to transform optical imaging technology as greatly as solid-state devices were transforming electronics. A CCD consists of a grid of tiny light-collecting "buckets"—called "pixels—at the surface of a silicon wafer. When a photon strikes a pixel, it frees an electron, which remains trapped in the bucket by the voltage applied to it. More photons eject more electrons, which remain trapped during an exposure. At the end of the exposure, the CCD

[13] Hagen, H.-J., *et al.* 1995, *Astron. & Astrophys. Suppl.*, **111**, 195, "The Hamburg Quasar Survey. I. Schmidt Observations and Plate Digitization..

[14] Wisotzki, L., *et al.* 1996, *Astron. & Astrophys. Suppl.*, **115**, 227, "The Hamburg / ESO Survey for Bright Quasars. I. Survey Design and Candidate Selection Procedures."

[15] "Integrated Circuits," http://www.pbs.org/transistor/background/events/icinv.html; accessed 6 May 2014.

[16] "Charge-Coupled Device," http://en.wikipedia.org/wiki/Charge-coupled_device; accessed 23 June 2013.

is "read out" by changing the voltages across the pixels to transfer the packets of electrons in sequence out of the CCD array and into storage for further processing. A CCD thus produces directly a digitized image, which is the basis for digital cameras. Of course, astronomers immediately adopted CCD detectors as well and began to employ them in a variety of applications.

The most extensive sky survey to date—and the most productive in discovering white dwarfs—is the Sloan Digital Sky Survey (SDSS), begun in the late 1990s.[17] Using a dedicated 2.5-m telescope at Apache Point in New Mexico, the SDSS employs a specially designed CCD camera with more than 100 million pixels to carry out a deep, systematic digital survey of the northern sky to a very faint limiting magnitude of approximately $V = +23$.

Although the main purpose of the survey was to identify candidate high-redshift objects, it is also highly effective in identifying potential white dwarfs. The SDSS began taking data in the year 2000 and has made its treasure trove of electronic data available to all astronomers by means of annual data releases. From more than 500 million objects, sophisticated data processing enables the selection of candidate objects likely to be stars—including white dwarfs—distant galaxies, and quasars. Digital spectra are subsequently obtained for each candidate object. As of 2006, the SDSS had produced a catalog of 9,316 spectroscopically confirmed white dwarfs, of which 6,000 had been unknown before the survey.

---

[17] The Sloan Digital Sky Survey is described in http://en.wikipedia.org/wiki/Sloan_Digital_Sky_Survey; accessed 19 May 2012; see also Gunn, J. E., and Knapp, G. R., in Soifer (1993), p. 267, "The Sloan Digital Sky Survey."

8. The Peculiar Spectra of White Dwarfs

Once a white dwarf candidate has been found, astronomers must obtain its spectrum and then analyze it in detail to determine its physical properties. Acquiring a spectrum of any star requires the use of a prism or diffraction grating to disperse (spread out) the starlight over all wavelengths for which there is significant radiant flux. Because white dwarfs are such faint stars, however, much less light is available to spread across the spectrum than is the case for ordinary Main Sequence stars. Although some white dwarf candidates identified in early surveys were actually bright enough for astronomers to secure their spectra as soon as they were discovered, many more were not.

An early leader in the effort to obtain white dwarf spectra was Gerard P. Kuiper[1] (1905–1973; see Figure 8.1), born a tailor's son in the Netherlands. He received his B.Sc. in astronomy from Leiden University in 1927, where he rubbed elbows with many of the bright young Dutch physicists and astronomers of his generation. He carried out his graduate research on a problem concerning binary stars with Ejnar Hertzsprung, who had supervised Willem J. Luyten's dissertation a dozen years earlier. Completing his Ph.D. thesis in 1933, Kuiper continued in Luyten's footsteps by accepting a postdoctoral position at Lick Observatory in California. Two years later, again like Luyten, Kuiper also went on to the Harvard College Observatory. There he met and married his American wife in 1936, becoming a naturalized U.S. citizen in 1937.

Kuiper's next position was at the University of Chicago's Yerkes Observatory, which had been founded less than half a century

[1] Gerard P. Kuiper's life and scientific contributions are summarized in "Gerard Kuiper," http://en.wikipedia.org/wiki/Gerard_Kuiper; accessed 19 May 2012.

H.M. Van Horn, *Unlocking the Secrets of White Dwarf Stars*, Astronomers' Universe, DOI 10.1007/978-3-319-09369-7_8

FIGURE 8.1 Gerard P. Kuiper. Credit: AIP Emilio Segré Visual Archives, Physics Today Collection

earlier. Named for Chicago street-railway tycoon Charles T. Yerkes, the observatory was established in 1892, when University of Chicago President William Rainey Harper and astronomer George Ellery Hale persuaded Yerkes to fund the construction of a new observatory.[2] It was to have the "largest and best" telescope in the world, a 40-in. refractor, besting the 36-in. refractor at Lick Observatory in California, which was then the largest.

Already demonstrating the entrepreneurial spirit that was to be his hallmark, the 24-year-old Hale was then just 2 years past his graduation from the Massachusetts Institute of Technology (MIT). The lens for the new Yerkes telescope—made by Alvan Clark and Sons, the same company that had figured the lenses for the Great Refractor at Lick Observatory a decade before—was installed near the end of May 1897.

[2] The action of University of Chicago President Harper and astronomer George Ellery Hale in persuading Charles T. Yerkes to build a new observatory is summarized in http://www.brittannica.com/EBchecked/topic/252289/George-Ellery-Hale; accessed 22 May 2012.

In 1941, Kuiper published a list of 38 stars for which he had obtained spectra and had confirmed as white dwarfs.[3] Only two spectra had been obtained before 1920, all but one from the Lick Observatory. The rest had been obtained after 1930, more than half between 1939 and 1941. Nine were proper-motion stars identified as white dwarf candidates by Luyten, while others came from a variety of early studies of nearby stars.

Already by the time of this paper, Kuiper could state that "It is well known that white dwarf spectra have only a remote resemblance to spectra of ordinary stars." The majority of the stars he listed—the 27 he classed as "wA," meaning white dwarfs containing hydrogen absorption lines—displayed "only the Balmer lines $H\alpha$ to $H\eta$ ... strongly broadened by the Stark effect, and no other lines." The unusually broad lines—with widths measured in many tens to hundreds of Ångstroms, in contrast to widths of at most a few to a few tens of Ångstroms for Main Sequence stars—were consistent with stellar atmospheres having very high pressures, as expected for stars like white dwarfs with very high surface gravities. A few stars showed the H and K lines of ionized calcium (denoted as "Ca II" by astronomers, in contrast to "Ca I" for neutral calcium atoms)—"though less strongly than in an ordinary F star"—and Kuiper classified them as "wF." For 7 of the 38 stars, Kuiper found no lines at all to be visible on low-dispersion spectra, and he labeled them "Con,"meaning that they exhibited essentially continuous spectra, uninterrupted by absorption lines.

Kuiper spent most of his career in Chicago, leaving in 1960 to found the Lunar and Planetary Laboratory (LPL) at the University of Arizona. He remained the director of LPL until his death in 1973. At Arizona, he provided major assistance to the National Aeronautics and Space Administration (NASA) in selecting potential landing sites for the Apollo missions to the Moon.

Kuiper's older colleague, Willem J. Luyten, who by the 1940s was already making a name for himself with his proper-motion and color searches to identify white dwarf candidates, also found himself interested in their spectra. In 1945, Luyten introduced a system of spectral classification specifically tailored to white

[3] Kuiper, Gerard P. 1941, *Publ. Astron. Soc. Pacific*, **53**, 248, "List of Known White Dwarfs."

dwarfs.[4] The majority, which show a spectrum consisting solely of broad hydrogen absorption lines, he called "DA," differentiating them from Main Sequence A stars by the prefix "D." The white dwarfs that exhibited only helium lines he classified as "DB" or "DO," depending on the degree of ionization of helium and thus the stellar temperature. Those in which no absorption lines were visible he termed "DC," indicating that they had a continuous or featureless spectrum. By the time Luyten published his new white dwarf spectral classification system, 80 white dwarfs had been confirmed.

To conduct spectroscopic studies of the fainter white dwarfs, however, would require still-larger telescopes with still-larger collecting areas. The idea for a very large optical telescope was the brainchild of George Ellery Hale (1868–1938; see Figure 8.2), who proposed it in 1928. Born to a wealthy Chicago family, Hale had become interested in astronomy at an early age.[5] For his senior thesis at MIT, he designed a spectroheliograph to enable spectroscopic investigations of the Sun over narrow wavelength bands. This work—and the backyard observatory the then 20-year-old Hale had built at his home in Chicago—was what had brought him to the attention of University of Chicago President Harper, who hired him in 1892 to bring astronomy to Chicago. When Hale and Harper secured funding to build the Yerkes Observatory, Hale ensured that it was equipped with ample laboratory space. The new observatory opened in 1897, and it supported a full program of solar and stellar research.

Hale continued to be interested in ever-larger telescopes, however,[6] and he soon had his staff busy fabricating a 60-in. reflector. In 1908 it was installed at the Mount Wilson Observatory in California, which had been established by the Carnegie Institution of Washington, itself then recently set up by steel magnate

[4] Luyten (1971), p. 3, "The White Dwarfs."

[5] George Ellery Hale's early life through his appointment at Chicago is sketched in http://www.brittannica.com/EBchecked/topic/252289/George-Ellery-Hale; accessed 22 May 2012.

[6] The histories of the Mt. Wilson and Mt. Palomar Observatories are summarized in http://www.astro.caltech.edu/palomar/history.html; accessed 22 May 2012. This includes Hale's construction of a 60-in. reflector for the Mt. Wilson Observatory as well as the selection of the site for the observatory on Mt. Palomar for the 200-in. telescope and the planning and development of that instrument.

FIGURE 8.2 George Ellery Hale. Credit: The University of Chicago Yerkes Observatory, courtesy AIP Emilio Segré Visual Archives

and philanthropist Andrew Carnegie. Even before this telescope was installed, Hale was already planning a still-larger 100-in. reflector for Mount Wilson. Delayed by the technical challenges involved in casting and polishing such a large mirror and then by World War I, this telescope was finally installed in 1918.

In the early 1930s, Hale began to search for a better astronomical site than the one at Mount Wilson for the 200-in. telescope he had envisioned in his 1928 article. He eventually found one on Mount Palomar, about 100 miles southeast of the campus of the California Institute of Technology (Caltech) in Pasadena. The new observatory's first telescope was an 18-in. Schmidt telescope (which has since been retired). Built with a $6 million grant from the Rockefeller Foundation, the 200-in. (today called the 5-m Hale Telescope) employed a single, monolithic mirror. After an unsuccessful effort to cast a suitable mirror blank for the giant new telescope from fused quartz, Hale selected Corning Glass-Works in western New York State to cast it instead from a then new glass blend called Pyrex.

The choice of Pyrex proved auspicious, as it is much less sensitive to changes in temperature than ordinary glass. It is thus less subject to distortion and focus problems like those that had plagued the 100-in. telescope on Mount Wilson. After the first Pyrex mirror cracked during cooling (it was subsequently displayed at a museum adjacent to the factory in Corning), the second casting was successful. In 1936 it was transported by rail across the nation to Pasadena on a special, slow-moving train, taking 16 days to complete the journey. Grinding and polishing the mirror then began in the optics laboratory at Caltech. This process took 11 years to complete, in part because work was halted in 1941 when the United States entered World War II. Work was resumed in 1945 after the end of the war, when the scientists and engineers returned from war-related work.

In November 1947, the polished 40-ton mirror, traveling in a huge protective crate on a flat-bed truck, slowly made its way from Pasadena to Mount Palomar and was then carefully installed in the waiting telescope structure inside the dome. The new 200-in. telescope (see Figure 8.3) was formally dedicated in June 1948 and named in honor of George Ellery Hale, who had died a decade ear-

FIGURE 8.3 The 200-in. Hale Telescope on Mt. Palomar. Public domain image courtesy Wikimedia Commons (http://upload.wikimedia.org/wikipedia/commons/b/b4/HaleTelescope-MountPalomar.jpg)

lier, never seeing the final fruition of his great vision. Astronomer Edwin P. Hubble (1889–1953) took the first photographic exposure with the 200-in. telescope in January 1949, and the giant instrument was made available for regular use by Caltech and Carnegie Institution astronomers that October. It remained the largest and arguably the most productive telescope in the world for the next 45 years.

Caltech's Jesse L. Greenstein immediately took advantage of the new 200-in. telescope to obtain spectra of white dwarf candidates, which by then numbered in the thousands. Born in 1909 in New York City, Greenstein (1909–2002; see Figure 8.4) was inspired to pursue a career in science as an 8-year-old, when his grandfather gave him a small brass telescope as a toy.[7] The precocious boy soon began experimenting with a variety of optical and electronic devices, including a spectroscope that sparked a lifelong interest in determining the composition of materials. He entered

Figure 8.4 Jesse L. Greenstein. Credit: AIP Emilio Segré Visual Archives, Physics Today Collection

[7] Trimble, Virginia 2005, *Proc. Am. Phil. Soc.*, **149**, 94, "Jesse L. Greenstein."

Harvard University at age 16, receiving his Bachelor's degree in 1929 and his Master's degree a year later.

By this time, however, the nation was in the grip of the Great Depression, so Greenstein decided to join the family real estate and finance business instead of continuing with his scientific career. His passion for science would not be denied, however, and he returned to Harvard in 1934 and received his Ph.D. 3 years later.

Receiving a prestigious fellowship from the National Research Council, Greenstein began his postdoctoral career at the University of Chicago's Yerkes Observatory. During this period he took part in designing and using novel spectrographs, which he employed for a variety of astronomical studies, including one that resulted in the discovery that interstellar grains are made from common materials such as ices and silicates (rock-forming substances) as well as one that demonstrated that some stars contain a mixture of chemical elements considerably different from the Sun and Main Sequence stars.

When his 2-year fellowship ended in 1939, Greenstein joined the astrophysics faculty at the University of Chicago. During World War II he worked with other Yerkes astronomers in conducting military research in optical design.

In 1948, Greenstein was appointed the first professor of astronomy at Caltech. There he was tasked to organize a graduate program in astronomy to help the university take full advantage of the magnificent new 200-in. telescope at the Palomar Observatory. He ended up serving as the head of the astronomy department for 24 years, rapidly building it into one of the best in the world. In contrast to many astronomy departments of the day, Greenstein insisted that astrophysical training be strongly grounded in physics. He was perhaps one of the first astronomers to take radio astronomy seriously, starting the Caltech radio astronomy group and subsequently playing a role in the establishment of the National Radio Astronomy Observatory in Virginia.

Greenstein's wide-ranging research interests centered on the physics of astronomical objects generally and of white dwarfs in particular. He studied more than 500 white dwarfs and other faint stars, collaborating with Olin J. Eggen (1919–1998) and others to carry out these investigations. This work greatly increased the sample of confirmed white dwarfs and added substantially to astronomers' knowledge of their colors, sizes, brightnesses, and surface compositions. Greenstein also directed then second-year graduate student Virginia Trimble—herself now a distinguished professor of physics at the University of California at Irvine as well as a visiting professor of astronomy at the University of Maryland—in obtaining measurements of the gravitational redshifts predicted by Einstein's general theory of relativity for a large sample of white dwarfs. We shall return to this work in Chap. 15.

During the 1960s, astronomers recognized that white dwarfs could be divided into two main groups according to their atmospheric compositions. Spectroscopic investigations by then had shown that hydrogen is the dominant atmospheric constituent for the DA stars. Unlike ordinary, non-degenerate Main Sequence A stars, however, the temperatures of the DAs span a very large temperature range, covering more than an order of magnitude in effective temperature (T_{eff}). Conversely, in Main Sequence A stars, hydrogen absorption lines dominate the spectrum only for effective temperatures near 10,000 K.

At the hot end of the DA sequence, astronomers found the Balmer lines of hydrogen ("H lines," for short) to be weak and quite broad. In progressively cooler stars, the H lines grow stronger

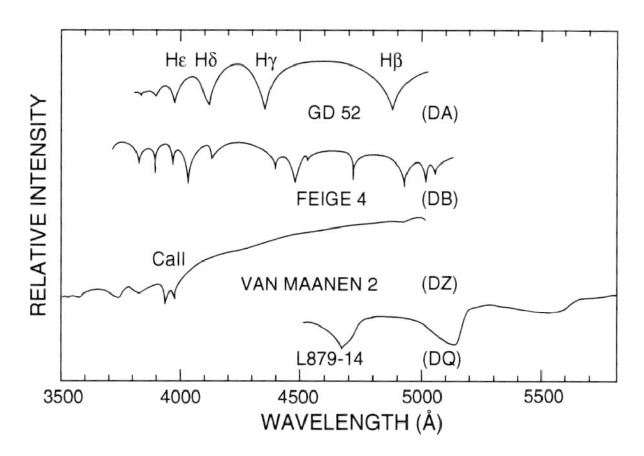

FIGURE 8.5 Sketches of a selection of white-dwarf spectra. Relative intensity increases *upward*, and wavelength increases to the *right*. The uppermost curve is the observed spectrum of the DA white dwarf GD 52. Several Balmer lines of hydrogen are labeled. The next curve shows the spectrum of the DB white dwarf Feige 4, in which a variety of neutral helium lines are visible. In the DZ white dwarf van Maanen 2, the doublet produced by ionized calcium at 3933 Å and 3968 Å is labeled. The DQ white dwarf L879–14 shows the Swan bands of molecular carbon at 4,737 Å and 5165 Å. Adapted from Wesemael *et al.* (1993)

until reaching a maximum around 12,500 K. GD 52, a DA star for which the spectrum is shown at the top of Figure 8.5, clearly exhibits these very strong hydrogen absorption lines, several of which are labeled in the figure. Below about 10,000 K, the Balmer lines rapidly become shallower and weaker, with the higher lines fading faster. Near 6,000 K only Hα remains strong with a marginally detectable Hβ line.

In contrast, the several different "non-DA" spectral types correspond to white dwarfs for which helium is believed to be the dominant atmospheric constituent. The non-DA sequence begins at the very highest temperatures—above 100,000 K—with the extremely hot PG 1159 stars. Next in temperature, beginning near 80,000 K, are stars with spectra dominated by absorption lines of ionized helium (He II). Greenstein classified such stars as DOs, in line with Main Sequence O stars, which also show He II absorption lines. Cooler DO stars (below about 50,000 K) show absorption by both He II and neutral helium (He I). Below 30,000 K, where He I lines dominate, the white dwarfs are classified as DBs. The DB star Feige 4, shown in Figure 8.5, has T_{eff} about 17,000 K and exhibits only strong, very broad absorption lines of neutral helium.

When the temperature falls below 11,000 K, the non-DA atmospheres become too cool to display optical He I lines, and if only helium is present, there are no spectral features to be seen at all. Such a star is classified as "DC," indicating a purely continuous spectrum. The very coolest of the DC stars, below about 5,000 K, are even too cool for the Balmer lines of hydrogen to be excited, and many of these very cool DCs may actually have hydrogen-rich atmospheres.

Due to the high transparency of neutral helium in atmospheres with effective temperatures less than about 9,000 K, the spectroscopist can often detect features due to trace abundances of carbon or heavier elements such as Ca, Mg, or Fe. Greenstein labeled white dwarfs displaying metallic features as spectral types DF and DG. Today they are termed "DZ" stars, using the symbol "Z" that astronomers employ for all the elements heavier than H and He, collectively. In these stars, lines due to ionized calcium (Ca II) are always seen, accompanied in some cases by lines of neutral iron and magnesium (Fe I and Mg I). A typical example is the

DZ star van Maanen 2 (Figure 8.5), which clearly displays the Ca II resonance doublet at 3,933 and 3,968 Å.

White dwarfs with traces of carbon in their atmosphere—originally classified as "λ4670" stars but now classified as white dwarfs of spectral class "DQ"—display the "Swan" band systems due to absorption by molecular C_2. In such a star, a convective envelope reaches deep into the helium surface layer of a former DB star until it touches and mixes upwards some of the underlying carbon core. Carbon is the trace element producing the spectral features in DQ stars such as L879-14, shown in Figure 8.5.[8]

Some of the warmer white dwarfs with traces of atmospheric carbon also exhibit absorption lines due to neutral C I. In cooler DQ stars only the molecular bands are seen, but the bands can be extremely weak, and in lower quality spectra, such stars may be classified as DCs.

For convenient reference, the primary characteristics of the white dwarf spectral classes are summarized below:

DA: Only H Balmer lines; no He I or metals present.
DB: He I lines; no H or metals present.
DC: Continuous spectrum, no lines deeper than 5 % in any part of the spectrum.
DO: He II lines strong; He I or H may be present.
DQ: Carbon features, either atomic or molecular, in any part of the spectrum.
DZ: Metal lines only; no H or He lines.

By the late 1980s, the number of confirmed white dwarfs had grown into the thousands, and the number of white dwarf suspects was many times larger. As of 1999, a catalog compiled by George P. McCook and Edward M. Sion of Villanova University contained more than 2,200 spectroscopically identified white dwarfs.[9] Many white dwarfs by this time had acquired numerous different identifying names, depending on how they had been discovered or studied—as members of binaries in long-established constellations, as proper-motions stars, as faint blue stars, as stars in spec-

[8] Adapted from Wesemael, F. *et al.* 1993, *Publ. Astron. Soc. Pacific*, **105**, 761, "An Atlas of Optical Spectra of White-Dwarf Stars."

[9] McCook, G. P. and Sion, E. M. 1999, *Astrophys. J. Suppl.*, **121**, 1, "A Catalog of Spectroscopically Identified White Dwarfs."

troscopic catalogs, as variable stars, and so on. The plethora of names had become confusing, so McCook and Sion devised a standardized naming system modeled after the one already in use for pulsars and quasars and essentially identical to the one employed by Richard Green in the PG survey of the late 1970s.

In this system, the star's name is preceded by the prefix "WD," followed by an abbreviation for the right ascension and declination. For example, the white dwarf Sirius B—also called α Canis Majoris B, because it is the companion (the "B") of the brightest star (the "α") in the constellation Canis Major (Latin for the "Big Dog")—is also called LFT0486 (because it is number 486 in a list compiled by Luyten in 1955 of stars with proper motions exceeding five-tenths of an arc second per year) and EG049 (object number 49 in a list of white-dwarf stars studied by Eggen and Greenstein). Because the astronomical coordinates of this star are right ascension 6 h 42 min 57 s and declination –16° 38.9′, the modern designation for Sirius B is WD0642-166. The WD numbers, taken from the McCook and Sion catalog, are listed in Appendix D of this book for most of the white dwarfs mentioned in this book.

9. Interlude: Crossing the Digital Divide

In the early twenty-first century, we have so much computing power at our fingertips—as evidenced in our smart phones and tablets and laptops—that it is hard to conceive of the difficulties scientists in the mid-twentieth century experienced in tackling some of their formidable computing challenges. Because of the complete transformation effected by the transition from calculations done by hand—essentially the entire period prior to World War II—to the advent of rapid machine calculations in the post-war period and especially after the early 1960s, it seems to this writer worthwhile to step back temporarily from our focus on white dwarfs to review briefly this remarkable change.

Until the middle of the twentieth century, modeling the evolution or the atmosphere of a star required laborious calculations by hand. A model of a single epoch in the evolution of a star might require the solution of equations to determine the temperature, pressure, luminosity, and radius at perhaps a hundred or more concentric, spherical mass shells. And the determination of the evolutionary development of a star might require hundreds of models. Similarly, the solution of the equations of radiative transfer and hydrostatic equilibrium in the atmosphere of a star might require the determination of the monochromatic intensity of radiation at hundreds of different frequencies, each such point requiring the determination of the wavelength-dependent opacity of stellar matter produced by hundreds of different energy levels in perhaps hundreds of different atoms and ions. Those calculations that were carried out employed mechanical adding machines, and the tedious work was often assigned to women "computers" or to graduate students.

H.M. Van Horn, *Unlocking the Secrets of White Dwarf Stars*,
Astronomers' Universe, DOI 10.1007/978-3-319-09369-7_9

Beginning around the middle of the century, however, such calculations began to be carried out by programmable digital computers. According to Jane Smiley, the biographer of physicist John Atanasoff (1903–1995), the all-electronic digital computer was invented by the then 34-year-old professor at Iowa State College in 1937.[1] In December of that year, Atanasoff had a brainstorm in which he envisioned the essential elements that would soon lead to the first electronic digital computer. His concept was, first, to use electronic logic circuits and, second, to employ binary numbers. The decision to perform calculations by simply turning electronic components on and off perfectly matched the binary number system, which has only the two digits, 0 and 1.

Within a few months, Atanasoff and his graduate student, Clifford Berry (1918–1963), "had constructed a working prototype at a cost of $650 ($450 to pay his assistant and $200 for materials)[2] .... [It] consisted of a breadboard-sized piece of wood on which Berry had built an electrical system of eleven vacuum tubes and fifty capacitors (or condensers, as they were called)."[3] Atanasoff had decided to use capacitors (devices that can store electrical energy even when not connected to an electrical power source) for memory storage in his machine. According to Atanasoff, the completed device "could just add and subtract the binary equivalents of decimal numbers having up to eight places,"[4] but it worked as he had conceived it.

After demonstrating the proof of principle with this breadboard device, Atanasoff and Berry began in January 1940 to construct a more sophisticated machine. Completed and ready for test in just a few months, the computer (called the ABC, for the Atanasoff-Berry Computer; see Figure 9.1), had a frame "seventy-four inches long, thirty-six inches deep, and about forty inches tall (including casters) ... ," according to Smiley.[5] It consisted of "a table with two levels, one about four inches off the floor that contained the boards holding the vacuum tubes and capacitors, which

[1] Smiley (2010), p. 1.

[2] *Ibid*, p. 3.

[3] *Ibid*, pp. 56 ff.

[4] *Ibid*.

[5] *Ibid*, p. 64.

FIGURE 9.1 A modern reconstruction of the Atanasoff-Berry Computer, or "ABC," originally built at Iowa State College in the early 1940s. Public domain image made available courtesy of Kerry Redshaw

stood upright and faced front, along with several other components, including two transformers and a power supply regulator.

> *Above that, at the back of the top table were two drums [for memory storage], each about eleven inches long and eight inches in diameter, several mechanisms for transposing binary numbers into and out of decimals, and a mechanism for charring holes in cards and feeding them back into the drums.*

The new machine could solve 29 linear equations with 29 unknowns, although it needed some 30 hours to do so and required systematic inputs by a human operator. Atanasoff described the ABC in detail in a 35-page manuscript that he completed in August 1940. Work on the ABC came to an abrupt halt in December 1941, however, when the United States entered World War II.

Additional impetus for the development of electronic computers came from the war effort. During the early years of the war, Herman Goldstine (1913–2004), a 28-year-old lieutenant in the U.S. Army with "a Phi Beta Kappa BA in mathematics and a Ph.D. in ballistics from the University of Chicago,"[6] was working at the Aberdeen Proving Ground in Maryland. He "was put in charge of ... [producing] firing tables [for artillery]," Smiley writes. "When

[6] *Ibid*, pp. 89 ff.

he took over, each table took a month to produce." At first he tried to "hire more women to do the computations ... [but] could find only a few ... ."[7]

About this same time, John Mauchly (1907–1980) and J. Presper Eckert (1919–1995) at the Moore School of Engineering of the University of Pennsylvania were also promoting the concept of an electronic digital computer[8]. Having heard of Mauchly and his computer ideas, Goldstine found him "and asked him about it ... . Mauchly submitted ... [a] proposal to Goldstine, and Goldstine sought authorization from the army to fund the project—conditions seemed dire and the army was desperate enough to grant $61,700 (the equivalent of $750,000 in 2010 dollars) to Mauchly and Eckert."[9]

Mauchly and Eckert's machine, an Electronic Numerical Integrator and Computer (called "ENIAC") "began to take shape in a large unused room at the... [University of Pennsylvania] in July 1943. At first the engineering team numbered twelve—Goldstine, Eckert, and Mauchly oversaw the general design (with Eckert in charge). Other members of the team were put in charge of individual components and, since the army was in desperate need of the firing tables, the Moore School team worked with seven-day-a-week dedication."[10]

"When it was completed in 1946, ENIAC was huge [see Figure 9.2]. It weighed twenty-seven tons, was eight feet long, eight feet high, and three feet deep. In addition to the 18,000 vacuum tubes, there were 7,200 diodes, 1,500 relays, 70,000 resistors, and 10,000 capacitors for memory storage. It required 150 kilowatts of power. ... Eckert said in 1989, 'We had a tube fail about every two days and we could locate the problem within 15 minutes.' ENIAC was not a programmable computer—its switches had to be set, and it had to be wired to perform its task; if the task changed, it had to be rewired and the switches reset. This could take weeks."[11]

---

[7] *Ibid.*

[8] Mauchly had previously visited Atanasoff and Berry at the Iowa State campus, and the question of who actually came up first with the key ideas for electronic digital computers was subsequently to become the focus of a bitterly contested lawsuit. The issue was finally decided in Atanasoff's favor only in 1973.

[9] Smiley (2010), p. 89 ff.

[10] *Ibid*, p. 93.

[11] *Ibid*, p. 95.

FIGURE 9.2 ENIAC, the Electronic Numerical Integrator and Computer built at the University of Pennsylvania in the mid 1940s. Public domain image courtesy Wikimedia Commons (http://commons/wikimedia.org/wiki/ENIAC)

According to Thomas J. Watson, Jr. (1914–1993), "When ENIAC was unveiled it created a huge stir ... It had no moving parts except for electrons flying at close to the speed of light inside its vacuum tubes. ... The most complicated problems of science and business often break down into simple steps of arithmetic and logic such as adding, subtracting, comparing, and making lists. But to amount to anything, these steps have to be repeated millions of times, and until the computer, no machine was fast enough. The quickest relay mechanism in our punch-card machines could only do four additions per second. Even the primitive electronic circuits of the ENIAC could do five thousand."[12]

Another spur to the development of electronic digital computers came from top-secret work on the development of the first atomic bomb at the Los Alamos laboratory in New Mexico. Richard P. Feynman (1918–1988), later to become a Nobel-Prize-winning

[12] Watson and Petre (1990), pp. 188ff.

physicist, was then working in the Theoretical Division at Los Alamos under Hans Bethe. Feynman recalled that in the early days of this work,[13] they "had to do lots of calculations, and we did them on Marchant calculating machines. ... You push .. [the numbered keys] and they multiply, divide, add, and so on ... They were mechanical gadgets, failing often, and they had to be sent back to the factory to be repaired. ...

"Anyway, we decided that the big problem—which was to figure out exactly what happened during the [atomic] bomb's implosion ... —required much more calculating than we were capable of. A clever fellow by the name of Stanley Frankel realized that it could possibly be done on IBM machines.[14] The IBM company had machines for business purposes, adding machines called tabulators for listing sums, and a multiplier that you put [punched] cards in and it would take two numbers from a card and multiply them. There were also collators and sorters and so on.

"Frankel figured out ... ," Feynman continued, that, if "we got enough of these machines in a room, we could take the cards and put them through a cycle. ... [This] was kind of a new thing then ... We had done things like this on adding machines. Usually you go one step across, doing everything yourself. But this was different—where you first go to the adder, then to the multiplier, then to the adder, and so on.

"Now we *always* were in a hurry. *Everything* we did, we tried to do as quickly as possible. In this particular case, we worked out all the numerical steps that the machines were supposed to do—multiply this, and then do this, and subtract that. ... [But] we didn't have any machine to test it on. So we set up this room with girls in it. Each one had a Marchant: one was the multiplier, another

[13]The discussion of wartime computational efforts at Los Alamos quotes extensively from Feynman (1986), pp. 108 ff, except as otherwise noted in the text.

[14]In the mid-1930s, IBM was a fairly new company," Smiley (2010), writes on p. 25, "the product of several mergers, but having its origins in the Tabulating Machine Company, which had been founded in 1896 by inventor Herman Hollerith – his first model had been used in the census of 1900. In 1911, several companies joined to form the CTR (Computing Tabulating Recording) Corporation, which offered a wide range of services to businesses ... Thomas J. Watson, Sr., had become president in 1915, and the name of the company was changed to International Business Machines in 1924. In 1928, IBM introduced the standard eighty-column punch card (the Hollerith card) that came to be familiar to students and secretaries for decades afterward. A 1931 model ... seemed exciting at the time—one astronomer declared himself thrilled just watching how quickly the machine went through its additions and subtractions."

was the adder. This one cubed—all she did was cube a number on an index card and send it on to the next girl.

> *We went through our cycle this way until we got all the bugs out. It turned out that the speed at which we were able to do it was a hell of a lot faster than the other way, where every single person did all the steps. We got speed with this system that was the predicted speed for the IBM machine. The only difference is that the IBM machines didn't get tired and could work three shifts. But the girls got tired after a while.*

"Fortunately the laboratory had already ordered IBM punch-card sorters to facilitate calculating the critical mass of odd-shaped bomb cores," according to Richard Rhodes in his excellent history of the making of the atomic bomb.[15] "The IBM equipment arrived early in April 1944 and the Theoretical Division immediately put it to good use running brute-force implosion numbers. Hydrodynamic problems, detailed and repetitious, were particularly adaptable to machine computation; the challenge apparently set [John] von Neumann thinking about how such machines might be improved."[16] Feynman was asked to take over the management of the IBM group, as the calculations were going very slowly. Realizing that "the young technical guys didn't know what they were working on," Feynman got permission to tell them. "The result was "*Complete* transformation! [They knew what they were doing and] *They* began to invent ways of doing it better. They improved the scheme. They worked at night. ... As a result, although it took them nine months to do three problems before, we did nine problems in *three* months, which is nearly ten times as fast."

In the meantime, mathematician John von Neumann (1903–1957)—invariably called "Johnny" by his colleagues—was commuting back and forth between Princeton and Los Alamos, where he also was helping with work on the atomic bomb. According to Rhodes, Lt. Goldstine recalled that sometime in the summer of 1944, " 'I was waiting for a train to Philadelphia on the railroad platform in Aberdeen [Maryland, the location of the US Army's Aberdeen Proving Ground] when along came von Neumann. ... When it became clear to von Neumann that I was concerned with

[15] Rhodes (1986), p. 544.

[16] *Ibid.*

the development of an electronic computer capable of 333 multiplications per second, ... the two of us went to Philadelphia so that von Neumann could see the ENIAC.' ... [von Neumann] soon abstracted from its crude vacuum-tube technology a logical system for manipulating and processing information, mathematical or otherwise."[17]

According to Smiley, "After meeting Goldstine, Eckert, and Mauchly, and chatting with Atanasoff at the ... [Naval Ordnance Laboratory, where he was then working] (and, no doubt, anyone else who seemed to know about computer theory)" von Neumann conceived a number of ideas for improvements in electronic digital computers. In June 1945, under von Neumann's direction, "Goldstine wrote a description of an idea for the second version of ENIAC. The paper was 101 pages long and was entitled 'first Draft of a Report of the EDVAC, by John von Neumann.' EDVAC stood for 'Electronic Discrete Variable Automatic Computer.' Partisans of von Neumann make the case that, as with everything else von Neumann did, he took the raw material of another man's ideas and immediately transcended it, or, ... [as one person said], 'Johnny grabbed other people's ideas, then by his clarity leapt five blocks ahead of them, and helped put them into practical effect.'"[18]

Thomas J. Watson, Jr., who succeeded his father in 1956 as the head of IBM, recalled that "Dad heard about Eckert and Mauchly late in the war, when the navy asked IBM to supply punch-card equipment to assist in getting data in and out of ENIAC ... . Eckert and Mauchly were talking about filing a patent, causing our lawyers to worry that IBM would have to pay big royalties if the idea of electronic computing ever went anywhere ... . [On] a gray day in March, ... we went to visit the ENIAC at the University of Pennsylvania. This was one of the first computers, a giant, primitive number-cruncher for solving scientific problems. It had just gone into operation, making a big name for its inventors, Presper Eckert and John Mauchly. ... I remember the ENIAC vividly. It was made up of what seemed like acres of vacuum tubes in metal stacks. The air was very hot, and I asked Eckert, a trim, urbane man, why that was. He explained, 'Because we are sharing

[17] Rhodes (1995), p. 250.

[18] Smiley (2010), p. 115.

this room with eighteen thousand vacuum tubes.' They hadn't air-conditioned it. I asked what the machine was doing and Eckert said, 'Computing ballistic trajectories.'... He explained that to make maximum use of a gun, you had to be able to calculate where its shell would be at every fraction of a second of its flight. This required a tremendous amount of computation, and ENIAC was doing it in a very short time—less time, in fact, than it would take an actual shell to reach its target. That impressed me ... . He said he and Mauchly were going to take the ENIAC patents and go into business."[19]

In June of 1947, Mauchly and Eckert filed patent applications for the ENIAC technology, which their agreement with the University of Pennsylvania allowed them to do. They set up a new company to build computers, with Mauchly as manager and Eckert overseeing construction. The new machine was to be called UNIVAC, for "UNIVersal Automatic Computer." By December of that year, the company had 36 employees. Two years later, the company ran into financial difficulties, and after an unsuccessful approach to IBM, it was sold to Remington Rand. "The first UNIVAC, which had been assembled on the second floor of the Eckert-Mauchly building, an old knitting factory, weighed 29,000 pounds and covered 380 square feet of floor space. It used 5,200 vacuum tubes (less than a third of the number in ENIAC) and consumed 125 kilowatts of electricity ... "[20]

By 1949, other "... more capacious, all-electronic successors to the ENIAC computer ... [were] being built at John von Neumann's direction at the Institute for Advanced Study in Princeton, [and] a copy [was] under construction at Los Alamos. Theoretical Division leader J. Carson Mark had lured Nicholas Metropolis from the University of Chicago in January 1949 to direct the computer project. As a joke ... the Chicago mathematician had named the Los Alamos computer MANIAC. [Russian émigré physicist George] Gamow decided that the pseudo-acronym stood for '*M*etropolis *a*nd *N*eumann *I*nvent *A*wful *C*ontraption')."[21]

[19]Watson and Petre (1990), p. 135.
[20]Smiley (2010), p. 150.
[21]Rhodes (1995), p. 383.

"The newspapers in the late 1940s were full of talk about laboratory computers with funny names like BINAC, SEAC, MANIAC, and JOHNNIAC [named after John Von Neumann]," Watson recalled. "Scientific conferences on computing and electronics were jammed. IBM had no plans to build such machines ... All of the new machines were cumbersome and enormously expensive, none was to be sold commercially, and for quite a while the ENIAC, the celebrated University of Pennsylvania machine ... was the only computer that actually worked."[22]

"We finally started producing computers after the outbreak of the Korean War," Watson continued.[23] "That was in June 1950 .... [IBM decided to] build a general-purpose scientific computer to work in *all* the defense applications ... . We gave our new computer a patriotic name: the Defense Calculator ... . Dad ... announced to shareholders at the annual meeting in April 1952 that IBM was building an electronic machine ... that was going to be rented and serviced along with our regular products. He gave it a number, the IBM 701, just like our other products ... "[24] The new machine "came off the production line in December 1952 .... Dad wanted to launch the 701 with all the usual IBM fanfare ... The ceremony was held in April, and one hundred fifty of the top scientists and leaders of American business showed up, including William Shockley, the inventor of the transistor, John von Neumann, the great computer theorist, General David Sarnoff, the head of RCA, and the heads of AT&T and General Electric. The guest of honor was J. Robert Oppenheimer, the brilliant physicist who led the scientific team that built the first atom bomb."[25] In 1954, IBM introduced another, smaller computer called the 650. "It was far less powerful than the Defense Calculator," Watson recalled, but much cheaper."[26] Consequently, it was snapped up by universities and businesses across the country.

During the 1950s, the fledgling computer industry was in the throes of the transition from vacuum-tube electronics to solid-

[22] Watson and Petre (1990), p. 188.

[23] *Ibid*, pp. 203 ff.

[24] *Ibid*, p. 207.

[25] *Ibid*, p. 229.

[26] *Ibid*, p. 243.

state devices, the transistor having been invented in 1947 at Bell Labs. "The transistor was obviously the wave of the future in electronics," Watson recalled. It "was faster than the vacuum tubes, generated less heat, and had great potential for miniaturization."[27] By the early 1960s, however, it had become clear that the future belonged not to discrete components like transistors but to "integrated circuits—computer chips that incorporate transistors, resistors, diodes, and so on, all in a single tiny unit. Nobody was using integrated circuits in computers yet, but ... [IBM's] System/360 design called for a lot of them ... This new line was named System/360—after the 360 degrees in a circle—because we intended it to encompass every need of every user in the business and scientific worlds ... "[28]

Another competitor in the burgeoning computer industry was the Control Data Corporation (CDC). "Control Data was the worst thorn in my side," Watson stated.[29] "It had been founded in 1957 by a team of electronics engineers who had worked together since the war, including a number of years at Remington Rand. Their leader was an entrepreneur named William Norris [1911–2006] and their top computer architect was Seymour Cray [1925–1996], a skinny, reclusive fellow who quickly became an industry legend. Norris's skill as a businessman and Cray's genius had made Control Data one of the industry's great success stories, growing in 6 years from nothing to annual sales of more than 60 million dollars. Their specialty was building big, ultrafast machines for the scientific market—what people now call supercomputers. These products appealed to the same clientele that gave the computer industry its impetus in the first place—weapons' labs, airplane and rocket manufacturers, elite universities—customers who were willing to fork over millions of dollars for the fastest state-of-the-art processors ... . Before Control Data came along, IBM had been at the top of the supercomputing game. Our flagship project in the late 1950s was a machine called STRETCH that grew out of a contract with the weapons lab at Los Alamos.[30]

---

[27] *Ibid*, p. 295 ff.

[28] *Ibid*, pp. 346 ff.

[29] *Ibid*, p. 382.

[30] *Ibid*.

> *In August 1963 came their [CDC's] bombshell: a machine called the 6600 that everyone recognized as a triumph of engineering. For seven million dollars, it delivered three times the power of STRETCH .... At that point the System/360 was the most advanced set of designs we had, and nothing in that whole product plan was even remotely comparable to the 6600 .... As it turned out, IBM was never able to beat the Cray design. Within two years, Control Data was back on its feet; scores of 6600s had been installed and its salesmen were dropping hints about even faster machines being built in Seymour Cray's lab ...* [31]

During and after the 1960s, the growing availability of electronic digital computers revolutionized all aspects of science and engineering. At universities across the globe and at national laboratories, "mainframe" electronic digital computers like the UNIVAC, the IBM 650 and System/360, and the CDC 6600 were immediately put to work solving many tedious calculations. This included modeling the structure and evolution as well as the atmospheres of stars. These models made it possible to interpret the growing numbers of observations.

[31] *Ibid*, pp. 383–384.

10. How to Make a White Dwarf

Inside every red giant there is a white dwarf trying to get out.

—Brian Warner

The catalyst for rapid progress in understanding the structure and evolution of stars was a computer algorithm devised by Louis G. Henyey (1910–1970) and his colleagues at the University of California at Berkeley in the late 1950s and early 1960s. The son of Hungarian immigrants,[1] Henyey received his Ph.D. in 1937 from the University of Chicago, with a thesis in astrophysical theory. A decade later he was appointed as a junior faculty member at Berkeley, where he established his own group to conduct research on stellar evolution.

The program Henyey and his colleagues devised[2] made possible for the first time truly automatic calculations of the evolution of stars, using the electronic digital computers that were by then proliferating across university campuses around the world. The basic idea underlying the "Henyey method," as it came to be called, is conceptually simple. Begin with an approximate numerical model for the distributions of pressure, temperature, radius, and luminosity at a number of concentric, spherical mass shells in a star at a given instant of time. Calculate corrections to each of the physical quantities so that the corrected model satisfies the stellar-structure equations better than the original approximation. Repeat the process until the residual errors are smaller than some pre-set limit. Save the resulting model as a satisfactorily accurate

[1] A brief biographical sketch of Henyey is given in http://en.wikipedia.org/wiki/Louis_G_Henyey.

[2] Henyey, L. G., Forbes, J. E., and Gould, N. L. 1964, *Astrophys. J.*, **139**, 306, "A New Method of Automatic Computation of Stellar Evolution."

H.M. Van Horn, *Unlocking the Secrets of White Dwarf Stars*, Astronomers' Universe, DOI 10.1007/978-3-319-09369-7_10

representation of the distributions of the physical quantities throughout the star at that instant of time. Now advance by a pre-determined step to the next instant in time, using the model just calculated as the initial approximation at the new time, and repeat the process until the new model has converged to the desired accuracy. Proceeding according to this program, the computer automatically calculates a sequence of models that trace out part of the evolution of the star.

The advantages of the Henyey method were so obvious that every astrophysicist interested in the theory of stellar evolution soon adopted some version of the new technique. One of the most successful of these theorists was Icko Iben, Jr. The son of a German immigrant who had become a professor of library administration at the University of Illinois,[3] Iben grew up in Champaign, Illinois. His father was a prolific author, with a focus on German culture and a special interest in Friesland, where he had lived. The son inherited his father's energy and productivity, but his own interests tended to physics rather than literature. In 1958 he received his Ph.D. from the University of Illinois for a thesis entitled "Higher Order Effects in Beta Decay: the Beta-Gamma Correlation and Time Reversal Invariance." Upon graduation, he accepted a teaching position at Williams College. Three years later, he moved to Caltech as a Senior Research Fellow, and his career in astrophysics began.

Four years before Iben went to Caltech, William A. Fowler (1911–1995), at Caltech's Kellogg Radiation Laboratory, together with Fred Hoyle (1915–2001) from Cambridge University in England, who was visiting Caltech at the time, and the husband-and-wife team of British astronomers E. Margaret Burbidge (then at Caltech) and Geoffrey R. Burbidge (1925–2010), then at the Mount Wilson and Palomar Observatories, had authored a ground-breaking paper describing the processes by which the chemical elements are synthesized in stars,[4] together with supporting astro-

[3]Information about Icko Iben, Sr., is given in http://www.library.uiuc.edu/archives/uasfa/3503025.pdf, accessed 7 July 2012, while a brief biographical sketch of Icko Iben, Jr., is given in http://www.amazon.com/Stellar-Evolution-Physics-Volume_Hardback/dp/110760253x, accessed 4 July 2012.

[4]Burbidge, E. M., Burbidge, G. R., Fowler, W. A., and Hoyle, F. 1957, *Revs. Mod. Phys.*, **29**, 548, "Synthesis of the Elements in Stars."

nomical evidence from observations of stellar spectra. Fowler evidently aroused Iben's interest in carrying out numerical calculations to track the detailed evolutionary processes by which nucleogenesis proceeds in stars. In his 3-year fellowship at Caltech, Iben did precisely that, using his own version of the Henyey method. In the process, he laid the foundation for a productive career in astrophysics, producing a steady stream of papers on all aspects of stellar evolution—both for single stars and for binaries—that would continue throughout his career. In 1964, Iben returned to teaching, this time at MIT, and in 1972 he went back to Champaign, when he was appointed to the faculty at the University of Illinois.

Iben was certainly not the only astrophysicist to apply some version of the Henyey method to stellar evolution calculations. Martin Schwarzschild (1912–1997) at Princeton University who, with his colleague Richard Härm, had been doing laborious hand calculations throughout the 1950s, immediately undertook the development of a similar code. In Germany, Rudolf Kippenhahn, Alfred Weigert (1927–1992), and their collaborators constructed yet another version of the Henyey code. So did Bohdan Paczynski (1940–2007) at the University of Warsaw in Poland. And back in North America, Pierre Demarque and his colleagues first at the David Dunlap Observatory of the University of Toronto and subsequently at Yale University also carried out stellar evolution calculations using another version. Each of these groups—and others—brought their own special expertise and interests to bear on the problem. In consequence, the general pattern of the evolution of stars of different masses rapidly came much more clearly into focus.

All these investigations had certain general features in common. To construct a numerical model of the physical structure of a star, one requires information about a number of physical properties of the matter of which it is composed. These include the so-called "equation of state" (the dependence of the pressure and entropy—a measure of the local heat content—on the local temperature, density, and chemical composition); the "opacity" (the quantity that controls the flow of radiant energy throughout the star, also as a function of the temperature, density, and composition); and nuclear reaction rates. Each of these physical properties depends in a complicated way on the degree of excitation and ionization of the atoms and ions present and on their interactions with each other. Before the advent of electronic digital computers, physicists and astrophysicists were forced to employ various simplifying assumptions in order to calculate approximate values for these quantities. It was difficult or impossible to test these calculations, because the conditions inside stars are so much more extreme than those available in terrestrial laboratories.

In the crash Manhattan Project to develop the atomic bomb during World War II, computations of material properties such as the equation of state and opacity were rapidly refined beyond any calculations that had come before. Conditions inside the imploding core of the bomb briefly reached temperatures and densities similar to those in the Sun. And the calculations needed to follow the progress of a thermonuclear explosion in a hydrogen bomb required material properties under conditions even more similar to those in the core of the Sun, which occur for a brief instant in the heart of the fireball. It is thus no surprise that the U. S. weapons laboratories became involved in calculations of the properties of matter under conditions like those inside stars—and had available the computing power necessary to deal with them as well as experimental facilities capable of subjecting them to increasingly rigorous tests.

In the 1960s and 1970s, Arthur N. Cox (1927–2013) and his colleagues at the Los Alamos National Laboratory published tables of the opacity for a variety of mixtures of chemical elements that were specifically designed for use in stellar evolution calculations.[5] The international Opacity Project (OP), led by Michael J. Seaton (1923–2007) of University College London, began in the 1980s with the goal of producing more accurate opacities based on improved atomic data.[6] And in the late 1980s and early 1990s, the OPAL opacity project, directed by Forrest J. Rogers (1938–2013) and Carlos A. Iglesias at the Lawrence Livermore National Laboratory (LLNL) in California, undertook a completely independent

[5] Cox, A. N., and Stewart, J. N., 1965, *Astrophys. J. Suppl.*, **11**, 22, "Radiative and Conductive Opacities for Eleven Astrophysical Mixtures;" see also Cox, A. N., and Stewart, J. N. 1969, *Astrophys. J. Suppl.*, **19**, 261, "Rosseland Opacity Tables for Population II Compositions."

[6] Seaton (1995).

program of opacity calculations based to the extent possible on fundamental physics.[7] The OP and OPAL projects estimated their results to be accurate to within a few percent.

The outcome of thc comprehensive stellar evolution calculations by Iben and others soon revealed the evolution of single stars in considerable detail. As we have seen in Chap. 5, whether or not they ultimately become white dwarfs, all stars begin their nuclear-active lives on the Main Sequence, burning hydrogen into helium at a central temperature of about 10 million Kelvin. This phase of central nuclear "H burning" is the longest phase of stellar evolution, extending over most of the lifetime of a star. For a one solar mass star, the H-burning lifetime is about 10 billion years. In contrast, the Main-Sequence lifetime of a 5 M_{Sun} star is only about 70 million years—because more massive stars have proportionately very much higher luminosities and therefore radiate away the energy generated by nuclear reactions very much faster, while stars that are less massive than about 0.8 M_{Sun} have H-burning lifetimes that exceed the age of the Galaxy and thus cannot yet have left the Main Sequence.

At the end of Chap. 5, we left our story of the evolution of a star after it had finished H burning, climbed up the red giant branch (RGB) in the H-R diagram, and reached core temperatures high enough to ignite He burning via the 3α reaction at the tip of the RGB. But how does a star go from there to become a white dwarf?

One of the first puzzles we encounter is that stars that will become white dwarfs have initial Main Sequence masses ranging up to about 8 M_{Sun}, while white dwarfs cannot have masses greater than the Chandrasekhar limit, about 1.4 M_{Sun}. We know that stars as massive as 8 M_{Sun} do become white dwarfs, because white dwarfs have been found in young star clusters, in which all of the stars are thought to have been formed at about the same time. An example is the white dwarf LB 1497 (WD 0349+247) in the nearby Pleiades cluster. The age of a star cluster is given by the H-burning lifetimes of the stars that are just finishing their Main Sequence evolution. For the Pleiades, this gives an age of roughly 70 million

[7] Iglesias, C. A., and Rogers, F. J., 1996, *Astrophys. J.*, **464**, 943, "Updated OPAL Opacities."

years.[8] Accordingly, the Main Sequence progenitor of this white dwarf must have had an initial mass greater than 5 M_{Sun}, and somewhere along the way it must have lost a very substantial fraction of its initial mass. When and how does this happen?

After helium ignition, a star again settles into an extended period of stable nuclear burning. Because the energy release from the He-burning reactions is less than that from H burning, and because this and later phases of stellar evolution occur at higher luminosities, all of these later evolutionary phases together add less than about 10 % to the lifetime of the star. During this phase of central helium burning, He is converted into C and O by the thermonuclear reactions in the core of the star, while the H-burning shell continues to gradually consume the stellar envelope, adding to the mass of He available as fuel.

At the end of the core-He-burning phase, a star thus contains an inner core composed predominantly of some mixture of carbon and oxygen, an overlying He layer, and an outer, still H-rich envelope. Exhaustion of the He-burning nuclear energy sources causes the C/O core to contract and the outer layers to expand, driving the star back up the giant branch in the H-R diagram for a second time. Because the star reaches a somewhat higher luminosity than it did on the first ascent of the red giant branch, after the end of core H burning on the Main Sequence, the evolutionary track corresponding to this phase of stellar evolution is called the "asymptotic giant branch" (AGB).

Because of the high degeneracy of the C/O core and of the He-burning region immediately above it, the star begins a prolonged series of "thermal pulses," during which quiescent H burning is interrupted by a series of violent He shell flashes (brief periods of runaway thermonuclear He burning), with successive flashes separated by about 100,000 years. Substantial mass loss does occur during these episodes.

Evolution on the AGB can be thought of as a race between the H and He shell sources—that respectively deplete the hydrogen envelope and add mass to the C/O core—and the mass-loss

[8] Salaris, M., *et al.* 2009, *Astrophys. J.*, **692**, 1013, "Semi-Empirical White Dwarf Initial-Final Mass Relationship: A Thorough Analysis of Systematic Uncertainties Due to Stellar Evolution Models."

mechanism, which removes the envelope from the outside. Empirically, in some cases more mass is lost from the outermost H layer than merely the amount required to quench the shell source and cause the remaining envelope to contract. This produces two sequences of post-AGB stars—with H-rich and H-poor surface layers, respectively.[9]

Uncertainties concerning the amount of mass loss, which have not yet been clarified by either observations or theory, preclude a completely quantitative description of the evolution of a star from the AGB to the immediate precursors of the white dwarfs. For these reasons, the relation between the initial Main Sequence mass of a star and the final white dwarf mass is still a subject of active research. A recent investigation[10] concluded that, for example, a star with initial mass of 1 M_{Sun} yields a C/O white dwarf with a final mass of about 0.5 M_{Sun}, while an 8 M_{Sun} star produces about a 1.4 M_{Sun} white dwarf.

So much matter is ejected from a star during the thermally pulsing AGB phase of evolution that it no longer can sustain a deep convective envelope, and it evolves rapidly to very high T_{eff} at an approximately constant luminosity L. Figure 10.1 sketches the evolution of stars from the Main Sequence through core He-burning and the subsequent thermally pulsing AGB phase to the final stage of cooling as a C/O white dwarf.[11]

The bend in the post-AGB tracks in this figure marks the transition from envelope contraction to the beginning of the diagonal white dwarf cooling track. Populated by the central stars of planetary nebulae (CSPN), it is traditionally called the "Harmon-Seaton" sequence, after the two scientists who first identified these stars in this penultimate phase of stellar evolution.[12] The CSPN are extremely hot stars, with effective temperatures T_{eff} up

---

[9] Iben, I., Jr., and Renzini, A. 1983, *Ann. Revs. Astron. Astrophys.*, **21**, 271, "Asymptotic Giant Branch Evolution and Beyond."

[10] Williams, K. A., Bolte, M., and Koester, D. 2009, *Astrophys. J.*, **693**, 355, "Probing the Lower Mass Limit for Supernova Progenitors and the High-Mass End of the Initial-Final Mass Relation from White Dwarfs in the Open Cluster M35 (NGC 2168)."

[11] Adapted from Iben, I., Jr. 1982, *Astrophys. J.*, **260**, 821, "Low Mass Asymptotic Giant Branch Evolution. I." and Iben, I., Jr. 1991, *Astrophys. J. Suppl*, **76**, 55, "Single and Binary Star Evolution."

[12] Harmon, R. J., and Seaton, M. J. 1964, *Astrophys. J.*, **140**, 824, "The Central Stars of Planetary Nebulae."

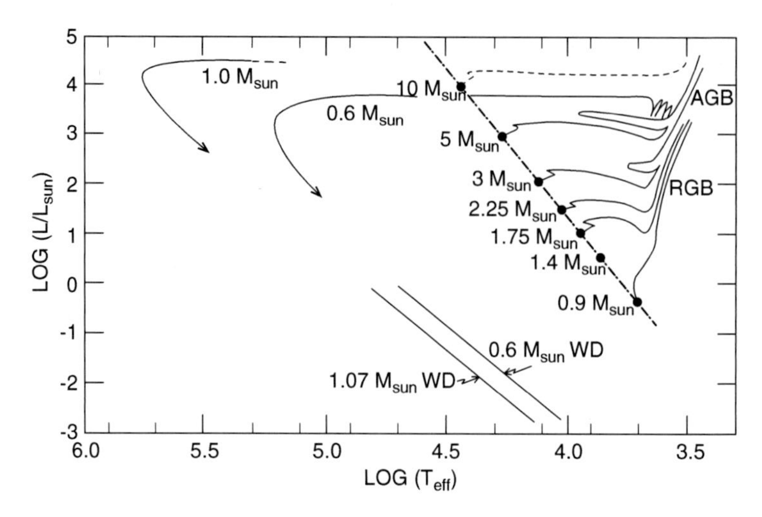

FIGURE 10.1 The evolution of stars that become white dwarfs, from the Main Sequence to the white-dwarf stage. The Zero-Age Main Sequence is shown as a *diagonal dot-dash line,* with the initial Main Sequence masses labeled. For stars less massive than about 2.3 M_{Sun}, the evolutionary tracks are sketched up the RGB, ending at the red-giant tip, where He burning is ignited. More massive stars loop back to the *blue* (higher effective temperatures, T_{eff}) during core He burning before ascending the AGB to much higher luminosities, *L*. On the AGB, stars undergo substantial mass loss through a series of thermal pulses. This transition stage of evolution is not shown. The final stages of two post-AGB tracks are sketched. For the 0.6 M_{Sun} remnant, the track begins near the AGB with a final series of thermal pulses. At the *bottom* of the figure, white-dwarf cooling tracks are shown for carbon white dwarfs of approximately 0.6 and 1.07 M_{Sun}. Adapted from (Iben 1982, 1991)

to 100,000 K or more, that radiate copiously in the ultraviolet. The UV radiation is absorbed by the surrounding thick shell of gas that was ejected from the star during mass-loss episodes earlier in its evolution and is then re-radiated at visual wavelengths, producing the beautiful, glowing nebulosities known as planetary nebulae (Figure 10.2).

Numerical stellar-evolution calculations[13] indicate that the He envelope of a CSPN is reduced to about 0.01 M_{Sun} by the time the He-burning shell is finally extinguished, while the H-burning shell rapidly reduces the mass of the surface H layer, ultimately

[13] He shell flashes in AGB stars and the residual masses of H and He in post-AGB stars are discussed by Iben and Renzini 1983, *op. cit.*

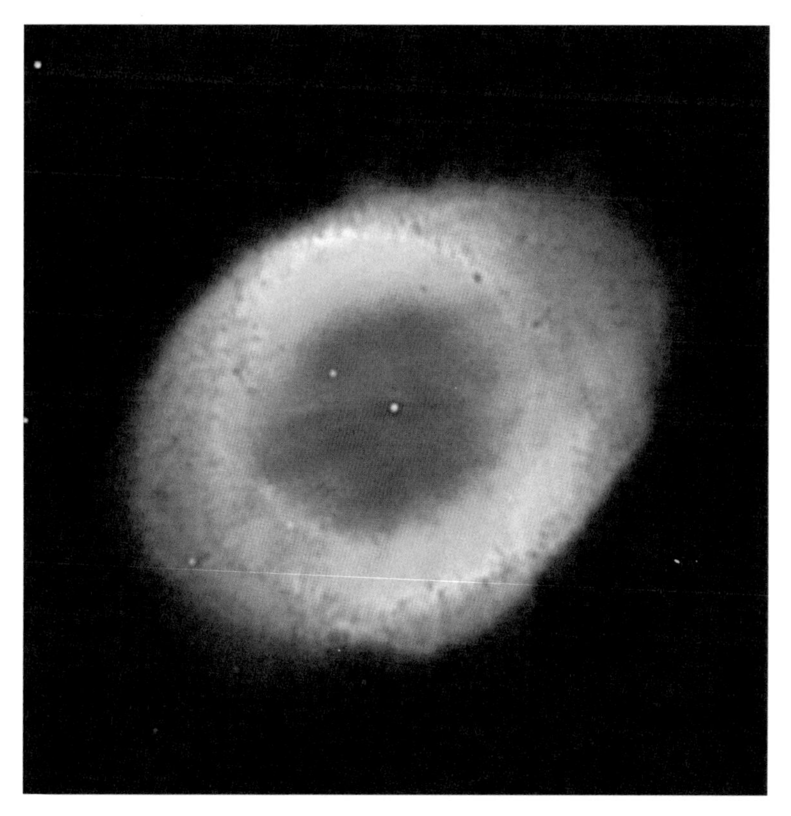

FIGURE 10.2 The Ring Nebula in the constellation Lyra. This beautiful nebulosity is a classic example of a planetary nebula. The white dot at the center is the very hot central star, on its way to becoming a white dwarf. Credit: C. F. Claver/WIYN/NOAO/NSF. Reproduced with permission

decreasing it to about 0.0001 M_{Sun}, when the base of the H layer is no longer hot enough to sustain nuclear burning. By this stage, most of the remaining mass of the star resides in its approximately 0.6 M_{Sun} C/O core, which is now sufficiently degenerate that it can no longer contract. Although the residual H- and He-layer masses at the surface of a white dwarf sound negligibly small, they actually play key roles in a number of observable effects, as we shall see in later chapters.

11. Diamonds in the Sky

Any Fool Can Make a White Dwarf!

—Icko Iben, Jr.

In 1961, Ed Salpeter (1925–2008) opened a new chapter in the investigation of white dwarfs. His goal was to improve the equation of state Chandrasekhar had developed and used in his pioneering white dwarf calculations in the 1930s. For simplicity, Chandrasekhar had assumed that white dwarfs are fully degenerate (that is, at the absolute zero of temperature), and he had neglected electrostatic interactions among the ions and electrons.

Salpeter retained the assumption that the ions and electrons were at zero temperature, but he included the full effects of the interactions. The most important correction he found was due to the electrostatic Coulomb interactions between the positive ions and the almost uniform, neutralizing background of negatively charged electrons. In consequence of these interactions, he pointed out, the ionic arrangement with the lowest energy at zero temperature is a body-centered cubic (bcc) lattice, rather than a gas or fluid, as had previously been assumed. About the same time, physicists D. A. Kirzhnits (1926–1998) and A. Abrikosov in the Soviet Union made the same point.[1]

Salpeter's calculations[2] revealed about a 10 % correction to Chandrasekhar's zero-temperature equation of state, with the actual values of the energy and pressure depending upon the charges of the ions. A bare $^{56}Fe_{26}$ iron nucleus, for example, with an ionic charge of $Z = 26$, produces a much larger correction than a

[1] Kirzhnits, D. A. 1960, *Soviet Phys.-JETP*, **11**, 365; see also Abrikosov, A. 1961, *Soviet Phys.-JETP*, **12**, 1254.

[2] Salpeter, E. E. 1961, *Astrophys. J.*, **134**, 669, "The Energy and Pressure of a Zero-Temperature Plasma."

H.M. Van Horn, *Unlocking the Secrets of White Dwarf Stars*, Astronomers' Universe, DOI 10.1007/978-3-319-09369-7_11

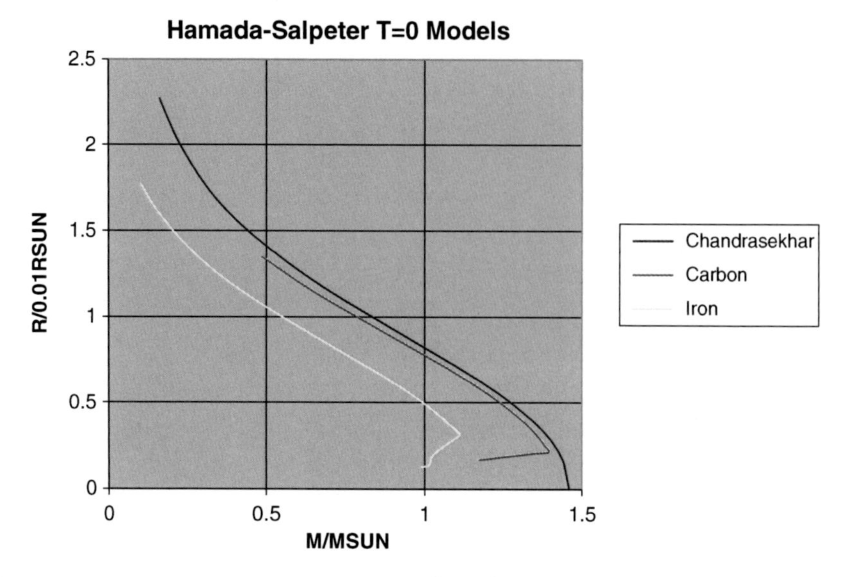

FIGURE 11.1 Zero-temperature Hamada-Salpeter models for white dwarfs. The uppermost curve in this figure gives the radii of Chandrasekhar's models (in units of one hundredth of the solar radius, a typical dimension for white dwarfs) plotted as functions of the stellar mass (in units of the mass of the Sun). The curve drops to zero radius at 1.46 M_{Sun}, the Chandrasekhar limiting mass. The next-highest curve shows the mass-radius relation for pure $^{12}C_6$ models constructed with Salpeter's equation of state, which includes the effects of electrostatic interactions among the electrons and ions. It reaches a maximum mass of 1.396 M_{Sun} at a finite radius. Models with yet-smaller radii appear beyond this point—at progressively smaller values of the stellar mass—but they are all unstable and cannot exist in nature. The lowermost curve shows the mass-radius relation for pure $^{56}Fe_{26}$ models. It reaches a maximum mass of only 1.112 M_{Sun}, and again models with smaller radii are unstable. Data from Hamada and Salpeter (1961)

$^{12}C_6$ nucleus or a $^{4}He_2$ nucleus. Consequently, the mass-radius relation for white dwarfs is not a single relation, as Chandrasekhar had found. Instead, there are slightly different relations for stars with different compositions. With his colleague T. Hamada, Salpeter calculated white dwarf models[3] with his new equation of state during a sabbatical year in Australia, and the mass-radius relations for these models are shown in Figure 11.1.

[3]Hamada, T., and Salpeter. E. E. 1961, *Astrophys. J.*, **134**, 669, "Models for Zero-Temperature Stars."

Six years after Salpeter published his results, Leon Mestel and Columbia University physicist Malvin A. Ruderman took the next step in improving our understanding of white dwarf interiors.[4] They realized that if the ions in the core of such a star actually do form a crystalline lattice, this would have significant consequences for the thermal properties of the star. Fifteen years earlier, Mestel had shown that the ions in the degenerate core of a white dwarf store essentially all of the thermal energy of the star.[5] In those calculations, he had assumed the ions to be an ideal gas, with heat stored as the kinetic energies associated with their random motions. If the ions instead form a crystalline solid, however, heat is stored in the form of vibrations of the ordered lattice.

In this case, experimental and theoretical studies had shown that the heat capacity of a solid at low temperatures drops off rapidly when the temperature falls below a characteristic value called the "Debye temperature,"[6] which depends on the density of the stellar matter. The rapidly falling heat content leads to accelerated cooling of white dwarfs at low luminosities, which has come to be called "Debye cooling." For a 0.6 M_{Sun} white dwarf, which has a central density of about 3.4×10^6 g cm^{-3}, the Debye temperature is about 3×10^6 K, comparable to the estimated central temperatures of the cooler white dwarfs. Because the Debye temperature is higher in more massive white dwarfs with greater central densities, they may actually cool rapidly enough to fade to invisibility in less than the age of the Milky Way Galaxy.

Where along the cooling track of a white dwarf star does crystallization occur, however? This question was tackled as an interesting intellectual exercise in 1966 by Stephen G. Brush, Harry L. Sahlin, and Edward Teller (1908–2003) at the Lawrence Livermore National Laboratory in California. Capitalizing on the availability of very high-powered (for the time) digital computers at the laboratory, they set out to calculate the thermodynamic properties of a collection of point ions in a uniform, neutralizing background—called a "one-component plasma" or OCP.[7]

[4] Mestel, L., and Ruderman, M. A. 1967, *Mon. Not. Roy. Astron. Soc.*, **136**, 27, "The Energy Content of a White Dwarf and Its Rate of Cooling."

[5] Mestel, L. 1952, *op cit.*

[6] Kittel and Kroemer (1980), pp. 102 ff.

[7] Brush, S. G., Sahlin, H. L., and Teller, E. 1966, *J. Chem. Phys.*, **45**, 2102, "A Monte Carlo Study of a One-Component Plasma."

In the course of this study, Brush and his colleagues found that the OCP spontaneously crystallizes at a critical value of a parameter they called Γ—the ratio of the characteristic energy of electrostatic interactions to the thermal energy—with lower temperatures (or higher densities) corresponding to larger values of Γ. In their calculations, the critical value at which the transition from the fluid to the crystalline phase occurs was $\Gamma \approx 125$. Subsequent, more accurate calculations of the fluid/solid phase transition in the OCP have given $\Gamma \approx 180$. For a 0.6 M_{Sun} ^{12}C white dwarf, this corresponds to a core temperature of about 4×10^6 K, while for a similar ^{16}O white dwarf it would be about 6×10^6 K. Again, these are similar to the estimated internal temperatures of the cooler white dwarfs, suggesting that crystallization—with all the attendant solid-state physics effects—is likely to occur during the latter cooling phases of white dwarfs. White dwarfs would not literally be "diamonds in the sky," because the crystal structure of diamond is not a bcc lattice, but they would be crystallizing bodies composed substantially of carbon.

I became interested in white dwarf stars at about this time. I had just completed my Ph.D. thesis at Cornell under Ed Salpeter's direction, investigating density-induced nuclear reactions (called "pycnonuclear" reactions from the Greek word "pyknos" meaning "dense) at very high densities, where the ions form a lattice. I was thus already interested in both astrophysics and solid-state physics. I had gone to the University of Rochester, in Rochester, New York, on a postdoctoral appointment with Malcolm P. Savedoff (Figure 11.2).

Not long after I arrived there in the fall of 1965—while Malcolm was away on a sabbatical year in the Netherlands—I became aware of the OCP calculations by Brush and his colleagues. I sent for a pre-publication copy of the Livermore report describing their work and studied it avidly when it arrived. My desk in the office I shared with another postdoc looked out across the Eastman Quadrangle in the heart of the campus, and I pondered the implications of the discovery of spontaneous crystallization of the OCP at sufficiently low temperatures as I watched the autumn days grow shorter on the Quad.

Two things seemed particularly clear. First, as physicists had long before shown, the transition from a fluid state, in which the

FIGURE 11.2 Malcolm P. Savedoff. Credit: AIP Emilio Segré Visual Archives, John Irwin Slide Collection

positive ions were randomly arranged, to a crystalline solid, in which the ions are located near the sites of a perfect lattice, cannot occur gradually and continuously but must take place discontinuously. This being so, the fluid-solid transition must be a first-order phase transition, which is accompanied by the release of latent heat, just as occurs when water freezes to form ice. Furthermore, from the results of the Livermore calculations, it was possible to calculate the approximate value of this latent heat for the crystallization of matter in a white dwarf, and it turned out to be a significant fraction of the thermal energy of the solid. Second, the heat content of a crystalline lattice was well known from solid-state physics to drop off rapidly when the temperature fell below the characteristic Debye temperature of the lattice. (I was unaware at the time of the work on Debye cooling then being done by Mestel and Ruderman.) It was clear that both these effects would significantly modify the cooling of white dwarfs from the calculations Leon Mestel had done in 1952. The release of latent heat would slow the cooling compared to Mestel's theory, while cooling would accelerate rapidly when the central temperature of the star fell below the Debye temperature.

Fired with enthusiasm for these new insights, I set about calculating the effects quantitatively. This was a period when the astronomers at Rochester were in the midst of a transition from doing numerical calculations by hand to doing them on digital computers. Savedoff was still then employing a woman "computer" to assist with tedious calculations—Stefania Zalitacz—and she carried out the numerical calculations I needed to determine the effects of crystallization on white dwarf cooling.

During the course of my postdoctoral years at Rochester, I made several trips to Cornell to consult with Ed Salpeter. The last half of the 90-mile drive follows the western shore of Cayuga Lake, one of the beautiful Finger Lakes in western New York, with the town of Ithaca and Cornell University located at its southernmost tip. During one of these trips I was excitedly telling Ed about my white dwarf crystallization ideas when he gently informed me that Mestel and Ruderman were working along very similar lines. Though crestfallen at learning this, I contacted Mal Ruderman upon my return to Rochester, and he kindly sent me a pre-publication copy of their paper. While they had indeed beaten me to the idea of Debye cooling, they had not included the release of latent heat on crystallization, so I did still have a contribution to make.

When Savedoff returned from his sabbatical year in Holland in the fall of 1966, he asked me to help him finish some numerical stellar evolution calculations begun by Samuel C. Vila for his Ph.D. thesis research under Malcolm's direction. Malcolm had previously acquired a copy of Martin Schwarzschild's stellar evolution code from Princeton University, and he and Vila had computed a number of sequences of models for cooling white dwarfs, assuming them to be composed entirely of iron.[8] This choice of composition was not intended to be realistic, but it enabled them to avoid dealing with nuclear reactions, since $^{56}Fe_{26}$ is the most stable nucleus in the Periodic Table.

This provided my first introduction to numerical calculations of stellar evolution, and I jumped into them with enthusiasm.

[8] Vila, S. C. (1965); see also Savedoff, M. P., Van Horn, H. M., and Vila, S. C. 1969, *Astrophys. J.*, **155**, 221, "Late Phases of Stellar Evolution. I. Pure Iron Stars."

The code Savedoff and Vila had developed employed a very simplified model for the non-degenerate surface layers of the star, essentially the same analytic model Mestel had used in his 1952 white dwarf cooling calculations. They also used a simplified equation of state that treated the ions as a non-interacting ideal gas immersed in a sea of non-interacting but partially degenerate electrons. The opacity was taken to be a combination of the ordinary radiative opacity—represented by a simplified analytic expression—and the "conductive opacity" originally calculated by Bob Marshak.[9]

Although Savedoff and Vila completely neglected any nuclear reactions and accompanying composition changes, they did include expressions for the neutrino energy-loss rates from the stellar core, and these proved to be important in the stage of evolution just prior to the beginning of the constant-radius cooling phase—near the bend in the evolutionary track that is populated by the central stars of planetary nebulae.

Marshak was still at Rochester when I arrived in 1965. I found him to be an energetic, open, friendly man, still greatly interested in the peculiar nature of white dwarfs from his thesis work with Bethe at Cornell almost three decades previously. He was highly admired by his colleagues on the Rochester faculty, not only because he was one of the university's top theorists in the developing field of high energy physics, but also because his orchestration of a decade-long series of "Rochester Conferences" on high-energy physics during the 1950s had attracted top scientists from around the world—even from the Soviet Union, despite this being the depths of the Cold War—and had quite literally put the Department of Physics and Astronomy at the University of Rochester "on the map" as one of the nation's then leading departments.

The next step in our developing understanding of white dwarf cooling began in 1969, with the arrival of Donald Quincy Lamb, Jr., (see Figure 11.3) as a graduate student in the Department of Physics and Astronomy at Rochester. Don had grown up in Manhattan, Kansas, obtained his Bachelor's degree in mathematics and physics from Rice University in 1967, and 2 years later received a Master's in theoretical high-energy physics from the University of

[9] See Chap. 6.

FIGURE 11.3 Donald Q. Lamb, Jr. Photo courtesy D. Q. Lamb, Jr. Reproduced with permission

Liverpool in England. He first showed up at the department office dressed in a jacket and tie and resembling those famous "Liverpuddlians" John, Paul, George, and Ringo. It did not take Don long, however, to adapt to the typical graduate-student "uniform" of jeans and tee shirt.

Don was initially undecided whether he wanted to continue in theoretical high-energy physics, which Marshak had firmly established as one of the strengths of the department, or whether he preferred to switch to theoretical nuclear physics or astrophysics. Ultimately, calculations of the evolution of crystallizing white dwarfs seemed the most interesting among the projects he was considering, and he began work on this topic under my nominal supervision. (I had been appointed an assistant professor in the department in 1967.) I actually regarded this more as a collaborative effort, since Don was not only a knowledgeable and capable young physicist but also a self starter and a hard worker.

Don began by constructing a computer code to calculate the equation of state accurately under the physical conditions expected throughout the stellar interior from the early, non-degenerate phases of evolution through increasing electron degeneracy, ion crystallization, and Debye cooling. He informed himself about, and made use of, the latest work by others for the several different components of this calculation, and the approach he adopted enabled him to compute the location of the crystallization phase transition for various chemical compositions. In particular, he computed "crystallization curves" showing the temperatures and densities at which the phase transition occurs for compositions of $^{4}He_{2}$, $^{12}C_{6}$, $^{16}O_{8}$, $^{12}Mg_{12}$, and $^{56}Fe_{26}$.

Lamb next turned to the stellar evolution calculations, beginning by taking the existing Rochester iron white dwarf program apart, streamlining the subroutines, and then reassembling them into a more efficient program. Then he set about improving the subroutines. He incorporated a vastly improved calculation for the non-degenerate surface layers of the star, making use of the work of another of his graduate student colleagues (described in Chap. 12). He also made major improvements in the handling of neutrino-loss rates, radiative opacities, and of course the equation of state, the latter handled by high-speed, accurate interpolations in pre-computed tables of values.

Using this improved code, Don calculated the evolution of a one solar mass pure $^{12}C_{6}$ model, starting from a non-degenerate initial model, following it through the phase dominated by copious neutrino emission from the core, and into the domain characterized by radiative cooling from the surface.[10] His calculations showed that crystallization and Debye cooling do have potentially observable consequences, making it possible in principle to test the theory by comparison with sufficiently accurate observational data.

After completing his thesis research and receiving his doctorate, Lamb accepted a 2 year postdoctoral appointment with Icko Iben at Illinois. He ultimately accepted a faculty appointment at the University of Chicago, where he remains a leading scientist to this day.

---

[10] Lamb, D. Q., Jr., (1974); see also Lamb, D. Q., and Van Horn, H. M. 1975, *Astrophys. J.*, **200**, 306, "Evolution of Crystallizing Pure ^{12}C White Dwarfs."

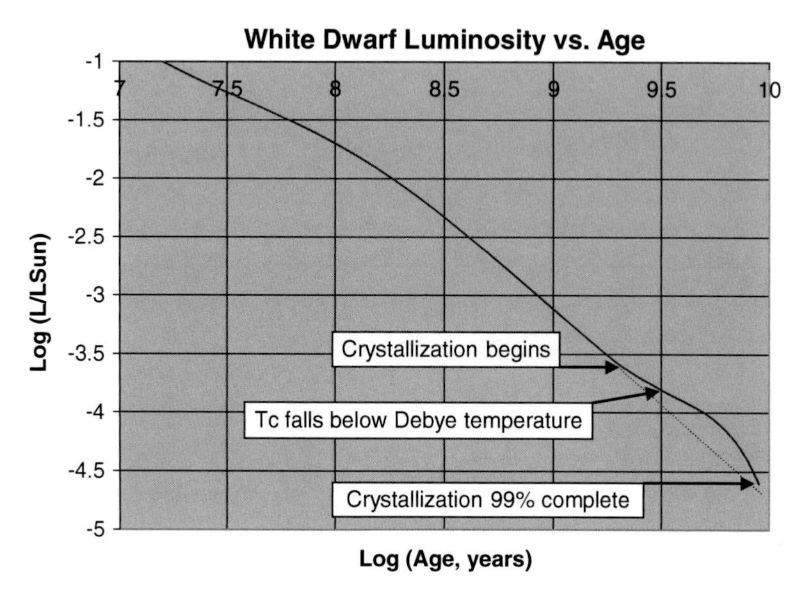

FIGURE 11.4 Cooling curve for white dwarf model that includes crystallization but not phase separation. The luminosity declines from 0.1 L_{Sun} at the top of the diagram to less than 3×10^{-5} L_{Sun} as the star's age increases from about 17 million years to nearly 10 billion years. The *solid curve* shows the actual model results, while the dotted line at low luminosities is just a continuation of the slope from higher luminosities. The center of the star begins to crystallize after about 2 billion years, slowing the rate of cooling as crystallization releases latent heat. At an age of about 3 billion years, the central temperature of the star falls below the Debye temperature, and the rate of cooling starts to increase as the heat capacity of the solid core drops. By the end of the calculation shown here 99 % of the mass of the white dwarf has crystallized. Data from Montgomery *et al.* (1999)

Lamb's white dwarf evolution code passed through many hands during the quarter century after his work, being continually refined in the process. By the end of the twentieth century much more realistic models had been computed. The cooling curve for one such model[11] is shown in Figure 11.4, illustrating the final evolution of a white dwarf with mass $M = 0.6\ M_{Sun}$ and surface He and H layer masses $10^{-2}\ M_{Sun}$ and $10^{-4}\ M_{Sun}$, respectively, as indicated by detailed calculations of pre-white dwarf phases of stellar evolution. The figure shows the decrease in luminosity of the

[11] Montgomery, M. H., Klumpe, E. W., Winget, D. E., and Wood, M. A. 1999, *Astrophys. J.*, **525**, 482, "Evolutionary Calculations of Phase Separation in Crystallizing White Dwarf Stars."

white dwarf as it cools over time. The vertical axis is the logarithm of the stellar luminosity in units of the luminosity of the Sun, and the horizontal axis is the logarithm of the stellar age in years. For comparison, the age of our Sun is about 4.65 billion years, corresponding to log (Age) = +9.67.

As can be seen in this figure, between about 10^8 and 10^9 years of cooling, the white dwarf luminosity declines in a manner quite similar to the result Mestel had obtained in his 1952 calculation. The dotted line extending to still lower luminosities is sketched to show what would happen if the star continued to follow this trajectory. However, by the time log (Age) ≈ +9.3—when log $(L/L_{Sun}) = -3.6$ and $\log(T_c) = 6.6$—crystallization begins at the center of the star, and the rate of cooling slows because the white dwarf requires additional time to radiate away the latent heat released during crystallization. The cooling track then lies above the (dotted) straight-line continuation of the original trajectory.

A few billion years later, the central temperature drops below the Debye temperature of 3.3×10^6 K at the white dwarf center. For a time, the continuing release of latent heat—as crystallization continues and the star slowly freezes from the center outward—competes with the decline of the heat capacity, as the temperature falls farther and farther below the Debye temperature. Ultimately, Debye cooling prevails, and the luminosity begins to drop more and more rapidly. By the end of the calculation, when the white dwarf is about 9 billion years old and log (L/L_{Sun}) less than –4.5, essentially the entire mass of the star has crystallized (apart from a very thin region near the stellar surface).

In 1979, David J. Stevenson suggested another possible energy source that may appear late in the cooling of a white dwarf, prolonging it still further. A New Zealander who became another of Ed Salpeter's Ph.D. students, Stevenson pointed out that the C/O mixture in a white dwarf core might undergo phase separation when the star begins to crystallize.[12] His Ph.D. thesis research,[13]

[12] Stevenson, D. J. 1980, *J. Phys. Colloq. (France)*, **41**, C2-61, "A Eutectic in Carbon-Oxygen White Dwarfs.

[13] Stevenson, D. J. 1975, *Phys. Rev.*, **B13**, 3999, "Thermodynamics and Phase Separation of Dense Fully-Ionized Hydrogen-Helium Fluid Mixtures;" see also Stevenson, D. J., and Salpeter, E. E. 1977, *Astrophys. J. Suppl.*, **35**, 221, "The Phase Diagram and Transport Properties for Hydrogen-Helium Fluid Planets."

published in 1976, concerned the effects of phase separation in H/He fluid mixtures in giant planets, which he showed to be capable of efficiently redistributing the chemical elements and simultaneously releasing significant amounts of gravitational potential energy. Three years later, he extended these ideas to C/O white dwarfs, where his calculations showed that a nearly pure O solid freezes out, leading to a substantial rearrangement of the C/O distribution in the core, accompanied by a significant release of gravitational potential energy. Stevenson has been a member of the faculty of the Division of Geological and Planetary Sciences at Caltech since 1980, where he is currently the Marvin L. Goldberger Professor of Planetary Science.

In 1983, Robert Mochkovitch confirmed that phase separation can be an efficient mechanism for chemical differentiation, with almost pure O "snowing" out to form the solid core.[14] The overlying, increasingly C-rich liquid remains well mixed by convection, driven by the resulting unstable concentration gradient. Because the crystallizing white dwarf must radiate away the gravitational energy released by this chemical segregation, the cooling time of the star is again increased, just as it was by the release of latent heat during crystallization. This affects the ages of the white dwarfs significantly, as the gravitational energy release may be appreciably larger than the latent heat.

As freezing continues, because the composition of the overlying fluid phase becomes increasingly carbon-enriched, the composition of the solid phase becomes progressively C-enriched as well. Depending on the nature of the phase diagram that describes the equilibrium between the fluid and solid phases, the rest of the star freezes either as a pure C crystal or as one with a fixed intermediate composition.

Several other groups have since studied the effects of C/O phase separation on the evolution of a cooling white dwarf, and it seems probable that it may lengthen the cooling times of white dwarfs by 2 Gyr or more by the time they have faded to luminosities as low as $\log(L/L_{\mathrm{Sun}}) = -4.5$.

---

[14] Mochkovitch, R. 1983, *Astron. & Astrophys.*, **122**, 212, "Freezing of a Carbon-Oxygen White Dwarf."

Even phase separation of minor species such as ^{22}Ne can produce further delays of as much as 2–3 Gyr, as pointed out in 1991 by Jordi Isern and his colleagues in Spain.[15] The actual abundance of ^{22}Ne in a white dwarf core is determined by the abundance of CNO elements in the material from which the white dwarf progenitor star was formed. This has varied throughout the history of our Galaxy. Stars that formed less than about 5 Gyr ago presumably had initial abundances similar to those in the Solar System. Early in the history of the Galaxy, however, the abundances of the heavy elements were much lower than they are today. Thus, the white dwarfs that originated from the earliest generations of stars may contain ^{22}Ne abundances orders of magnitude less than the abundances in white dwarfs that are just now beginning their cooling. The ages of the oldest white dwarfs are thus inextricably bound up with the history of chemical enrichment in the Galaxy. Whether or not this information can actually be extracted, however, remains to be determined in the future.

[15]Isern, J., Mochkovitch, R., Garcia-Berro, E., and Hernanz, M. 1991, *Astron. & Astrophys.*, **241**, L29, "The Role of the Minor Chemical Species in the Cooling of White Dwarfs."

12. The Envelope, Please!

In his seminal book on white dwarfs,[1] the late French astrophysicist Evry Schatzman (1920–2010; see Figure 12.1) first pointed out that cool white dwarfs must develop surface convection zones. Born to Romanian parents[2] who had emigrated to Palestine, Schatzman began his university studies in France in November 1939 just as World War II broke out. He fled the Nazi-occupied area of France after the German invasion, moving first to Lyon and then to the Haute-Provence Observatory. He was rumored to have been active in the French Resistance during the war.

When the conflict ended in 1945, he began work at the Centre National de la Recherche Scientifique (CNRS), receiving his doctorate a year later. He joined the faculty at the University of Paris in 1949 and remained there until moving to the Nice Observatory in 1976. He actively encouraged young people to develop interests in astronomy, and he was directly responsible for training some of the best young astrophysicists in France.

During the 1940s, Schatzman's research interests focused on white dwarf stars, where he made several important contributions. In his 1958 book, he investigated a number of problems in the theory of dense matter and explored the consequences for astrophysicists' understanding of white dwarfs. However, his prescient recognition that convection must develop in the envelopes of cooling white dwarfs was generally ignored at the time, perhaps because the observational data then available were insufficient to enable astronomers and astrophysicists to capitalize on it.

What Schatzman recognized was that when a white dwarf has cooled to the point at which the main chemical constituents in its surface layers can no longer remain fully ionized, a convection

[1] Schatzman (1958).

[2] Schatzman's early life is briefly summarized in http://en.wikipedia.org/wiki/%C3%89vry_Schatzman.

H.M. Van Horn, *Unlocking the Secrets of White Dwarf Stars*, Astronomers' Universe, DOI 10.1007/978-3-319-09369-7_12

FIGURE 12.1 Evry Schatzman in 1973. Credit: AIP Emilio Segré Visual Archives, John Irwin Slide Collection

zone must develop there. The reason is twofold. First, because incompletely ionized species possess bound states populated by electrons, additional processes—which are not available in a fully ionized plasma—become available to absorb radiation, so the opacity of the stellar matter rises. This makes it harder for radiation to escape, so the temperature gradient necessary to sustain a given flux of energy becomes steeper. At the same time, the heat capacity of the partial ionization zone increases. Both the energy necessary to ionize the atoms and that which can be absorbed as kinetic energies of the newly freed electrons contribute to this. And the increase in both the opacity and the heat capacity contribute to making a partial ionization zone unstable to convection.

In 1906, German astrophysicist Karl Schwarzschild had determined the conditions under which convection would spontaneously develop in a star.[3] He began by considering a layer of

[3] Schwarzschild (1958), pp. 44 ff.

matter in hydrostatic equilibrium under gravity. Assuming that a temperature gradient spans the layer, causing a steady flow of heat through it, he imagined a small bubble of matter to be displaced vertically from its lower boundary. If the bubble does not exchange heat with its surroundings—that is, if it undergoes what is called an "adiabatic" displacement—then to balance the internal pressure in the bubble with that in its surroundings (in other words, to maintain hydrostatic equilibrium), the bubble must expand, since the force of gravity across the mass layer causes the pressure at the bottom to be greater than that at the top.

Now, the density of matter depends on the local pressure and temperature, as expressed through the equation of state. The density change inside the bubble, produced by the assumed adiabatic displacement, is thus not necessarily the same as the density change in the surroundings, produced by hydrostatic equilibrium under gravity combined with the given temperature gradient. Consequently, Schwarzschild reasoned, if the density in the bubble is lower than the density in the surrounding matter, the bubble will be buoyant and will continue to rise. In other words, convection will begin spontaneously. This can be expressed as a condition involving the given temperature gradient: If it is steeper than the adiabatic temperature gradient—that is, if the temperature in the displaced bubble is higher than that of its surroundings, so that the density in the bubble is lower than its the surroundings—then the mass layer is convectively unstable.

What are the consequences for a star when it develops a convection zone? For one thing, studies have shown that convection in stars is generally very turbulent, like a boiling pot. For another, convection tends to be much more efficient at transporting heat than is radiation. Consequently, when a layer inside a star does become unstable, the convective motions carry essentially all the heat flow, and the temperature gradient remains fixed at essentially the adiabatic gradient. Convection also mixes material efficiently, so that the distribution of the chemical elements remains uniform throughout the convection zone.

A semi-empirical theory for the transport of energy by convection—called the "mixing-length theory"—was first developed in 1925 by the German aerodynamicist Ludwig Prandtl (1953–1975).[4]

---

[4] Ibid, pp. 47 ff.

The basic idea is that a buoyant convective bubble will rise though a prescribed distance—the mixing length *l*—and then mix completely with its surroundings. The excess heat carried by the bubble is released at that point. The mixing length itself is not determined by the theory but must be provided separately.

A natural characteristic length scale is the local pressure scale height—the distance given by dividing the ambient pressure by the pressure gradient—and a conventional assumption is that the mixing length is some fixed fraction of the pressure scale height. The numerical value of this fraction is chosen so that the resulting convective efficiency matches some empirical criterion. Of course, this is not a particularly satisfactory state of affairs, but with a few exceptions the results generally are not terribly sensitive to the particular choice made (We discuss in Chap. 17 one way to choose the mixing length that seems to work well for white dwarfs.)

Prandtl's theory was adapted to the deep interiors of stars in 1945 by German astronomer Ludwig Biermann (1907–1986). And in 1958, German-born astronomer Erika Böhm-Vitense (see Figure 12.2) extended the theory to conditions appropriate to stellar atmospheres.[5] Born in 1923, Erika Vitense was a young teenager at the start of World War II. After the war, she studied at Kiel University with the renowned German astrophysicist Albrecht Unsöld. There she met Karl-Heinz Böhm (1923–2014; see Figure 12.3), himself another promising Unsöld student, and the two were married in 1953. Beginning the following year, the Böhms became frequent visitors to the United States, and they both ultimately accepted faculty appointments in the Department of Astronomy at the University of Washington (Seattle). Individually and jointly, the Böhms have made substantial advances in a broad range of astronomical and astrophysical subjects during their productive careers.

Böhm-Vitense's version of the mixing-length theory is specifically intended for use in the surface layers of a star, where the convective elements begin to become transparent to radiation. This leads to energy loss from rising bubbles, a rapid decrease in the efficiency of convective energy transport, and ultimately—as

[5] Böhm-Vitense, E. 1958, *Zeitschrift für Astrophys.*, **46**, 108, "Über die Wasserstoffkonvektionszone in Sternen verschiedener Effektivtemperatur und Leuchtkräfte. Mit 5 Textabbildungen.".

Figure 12.2 Erika Böhm-Vitense. Credit: Department of Physics, University of Illinois at Urbana-Champaign, courtesy AIP Emilio Segré Visual Archives

Figure 12.3 Karl-Heinz Böhm in 1973. Credit: AIP Emilio Segré Visual Archives, John Irwin Slide Collection

the opacity decreases near the stellar surface—a return to radiative transfer as the dominant mode of energy transport. In particular, her theory is appropriate for the surface convection zones in white dwarfs, the upper parts of which extend up into the optically thin (*i.e.*, increasingly transparent) regions of the stellar atmospheres. Given the value of the mixing length, Böhm-Vitense's version of the theory allows one to calculate such quantities as the average velocity of the convective elements, the (slight) excess of the temperature gradient above the adiabatic gradient, and the heat flux carried by convection.

A decade after Schatzman's recognition of the potential importance of convection in cool white dwarfs and Erika Böhm-Vitense's development of a form of the mixing-length theory suitable for stellar atmospheres, her husband, Karl-Heinz Böhm, undertook the construction of a model for the cool white dwarf van Maanen 2. He assumed the surface layers to be 98 % pure helium, took the effective temperature to be $T_{\rm eff} = 5{,}790$ K, comparable to the effective temperature of the Sun, and used for the surface gravity the value $g = 10^8$ cm s^{-2}.[6] These values of g and $T_{\rm eff}$ for van Maanen 2 had been determined in 1960 by the German white dwarf expert Volker Weidemann (1924–2012). For this model, Böhm found the core temperature of the white dwarf to be reduced by about a factor of four—a huge change!—in comparison with a model that assumed energy transport through the envelope to be purely radiative. And this large decrease in the computed core temperature led to a correspondingly dramatic reduction in the calculated cooling time for the white dwarf. This was the first time anyone had realized how significantly convective energy transport can affect the thermal structure in the surface layers of white dwarfs, with important consequences for the relation between the white dwarf luminosity and core temperature and thus the white dwarf cooling time. Astronomers subsequently learned that these thin surface convection zones also play key roles in other aspects of white dwarfs, as we shall see in later chapters.

[6] This large a surface gravity is appropriate for a white dwarf star, with a mass about half that of the Sun and radius a hundred times smaller than the Sun's. This large value is about 3,000 times greater than the surface gravity of the Sun and more than 100,000 times greater than that of Earth.

Böhm's discovery of the importance of convection in cool white dwarfs stimulated my own interest in this problem, and I began to correspond with him about it. I also undertook my own simplified calculations of ionization equilibrium in white dwarf envelopes and constructed envelope models for a range of white dwarf masses, luminosities, and envelope compositions. All the models proved to be convectively unstable at the surface, and at sufficiently low temperatures the base of the convection zone even extended into the degenerate core. Assuming the core to be isothermal, as in Mestel's theory of white-dwarf cooling, this led directly to a rapid decline in the core temperature with continuing decrease in the stellar luminosity, directly confirming Böhm's findings. The story proved to be more complicated than our original simplified models had indicated, however.

The discovery that convection in the surface layers of white dwarfs can have such a dramatic effect on the temperature at the degeneracy boundary showed clearly that a more accurate treatment of convective energy transfer through the white dwarf surface layers was warranted. Because this required an accurate treatment of the equation of state and opacity in these layers, I began a research program to incorporate such improvements in white dwarf envelope calculations. Both at the University of Washington and the University of Rochester, graduate students were involved in these projects, so I explored with Karl-Heinz possibilities for dividing the work so that his students and mine would have projects that complemented, rather than interfered, with each other. Karl-Heinz proved very amenable to this, and it was easy to find a mutually agreeable approach. Karl-Heinz and his students focused on the upper part of the convection zone and the stellar atmosphere, ultimately developing a sophisticated method for the very difficult task of incorporating convection into calculations of the transfer of radiation through the model atmosphere of a white dwarf and using the results to determine greatly improved values for the observed atmospheric abundances of the elements in cool white dwarfs. In contrast, my students and I concentrated on the material properties in the deeper parts of the convection zones, incorporating electron degeneracy, Coulomb interactions, and other effects into the equation of state and opacity, and computing detailed models for white dwarf envelopes.

FIGURE 12.4 Gilles Fontaine at the Université de Montréal in 1999. Image courtesy G. Fontaine. Reproduced with permission

The student who became principally involved in this work at the University of Rochester was a very capable young French-Canadian physicist named Gilles Fontaine (see Figure 12.4). Born in 1948 in the province of Québec, he earned his Bachelor of Science degree in physics from l'Université Laval in 1969 and began his graduate studies at the University of Rochester that fall. He initially intended to study quantum optics, in which the Rochester department has long had a strong program, although he also had a considerable interest in astronomy.

A tall, thin young man with an unruly shock of hair and an engaging smile, Gilles rapidly proved himself an extremely capable researcher. Nor was he one-sided and single-minded. He had been an avid hockey player from a very young age, and he rapidly became involved with other graduate students in organizing pick-up games. Though I never saw him play, he developed a reputation as a fierce and wily competitor. He suffered a severe leg injury during one of these games, which has left him with a slight limp to this day.

Gilles was also recently married when he came to Rochester. His wife, Francine, knew no English when they first arrived, although Gilles was fluent in both languages. With Gilles away all day at the university, Francine was left to her own devices, and she later told me that she spent it learning English by watching

American soap operas on television. I learned later that Gilles was also an accomplished musician, partial to French-Canadian folk songs, and that he loved a good party.

Gilles decided to work with me on his Ph.D. thesis research, and we set out to investigate systematically the properties of convection in the envelopes of white dwarfs. The problem of determining ionization equilibrium in the surface layers of these stars is complicated by interactions among the charged particles, collectively termed "non-ideal effects" because they represent departures from the treatment of the ions and electrons as ideal gases. In addition, at sufficiently high densities the atoms and ions are squeezed together too closely to enable electrons to remain bound, and matter is said to be "pressure ionized." It remains a challenge to this day to determine exactly how and at what pressure or density this occurs. Approximate estimates yield a density of about 3 g cm^{-3} for H to become "pressure-ionized"—also termed the "metallization" of hydrogen, because the newly freed electrons conduct electricity well, as in a metal—and perhaps 85 g cm^{-3} for He.

About the time Gilles was beginning his Ph.D. thesis research, we became aware that Harold C. ("Hal") Graboske, Jr., and his colleagues at the Lawrence Livermore National Laboratory (LLNL) were working on detailed numerical computations of the properties of matter at temperatures and densities similar to those expected for the envelopes of white dwarfs. We contacted Hal, and he generously agreed to provide us with the equation of state he and his LLNL colleagues had calculated.

The approach the LLNL scientists had adopted was to construct a model for the "free energy" F (a well-defined thermodynamic quantity) for the mixture of atoms, ions, and electrons in a partially ionized plasma and then to determine the thermodynamic properties numerically from this function. The model they employed included contributions from partially degenerate electrons, Coulomb interactions among the various charged particles, and non-Coulombic interactions due to the finite sizes of the incompletely ionized heavy particles.

The resulting tabular equation of state included all of the thermodynamic quantities as functions of the temperature and density. This approach, in which individual ionization states of atoms and ions are treated as separate species, is called the "chemical picture." Most of the detailed equation-of-state calculations that have been performed for applications to the envelopes of the white dwarfs are of this type. Tabular equations of state for hydrogen, helium, and other elements were published in the late 1970s by Fontaine, Graboske, and myself[7] in the United States and by G. Magni and Italo Mazzitelli[8] in Italy.

In the alternative "physical picture," only the electrons and atomic nuclei are treated as separate species. All of the thermodynamic properties of the system can then be obtained from the solution of the quantum many-body problem. This method was extensively developed by Forrest Rogers (1938–2013) and his collaborators at LLNL beginning in the early 1980s and provides the basis for the so-called "OPAL" opacity calculations.

Since the OPAL opacities were not available when Gilles carried out his model-envelope calculations, he employed the tabular radiative opacities that had then been recently published by Arthur N. Cox and his collaborators at the Los Alamos National Laboratory (LANL).[9] In principle, these opacity calculations could have provided the thermodynamic quantities Gilles needed as a byproduct, but in practice, the LANL scientists had not included them in the output of their calculations.

Gilles began his white dwarf envelope calculations[10] starting with the very low-density atmosphere of the star, where radiation escapes freely into space. Beginning at the lowest densities, he worked his way inward, testing the model at each step to see where convection first set in. Then he continued the calculation inward with convection taken fully into account. Gilles did three separate sets of calculations, one set each for a composition of nearly pure H, nearly pure He, and nearly pure C. Convection only occurred in models cool enough for the stellar surface to no longer be completely ionized. Gilles confirmed that a convection zone begins to

---

[7] Fontaine, G., Graboske, H. C., Jr., and Van Horn, H. M. 1977, *Astrophys. J. Suppl.*, **35**, 293, "Equations of State for Stellar Partial Ionization Zones.".

[8] Magni, G., and Mazzitelli, I. 1979, *Astron. & Astrophys.*, **72**, 134, "Thermodynamic Properties and Equations of State for Hydrogen and Helium in Stellar Conditions.".

[9] Cox, A. N., and Stewart, J. N. 1969, *Astrophys. J. Suppl.*, **19**, 261, "Rosseland Opacity Tables for Population II Compositions."

[10] See Fontaine (1973) for details.

develop at the surface of a white dwarf when the effective temperature of the star falls below the value needed to maintain the dominant atmospheric constituent in a fully ionized state. Because of the high surface gravities of the white dwarfs, this occurs at relatively high T_{eff}: about 25,000 K for DB (He-rich) white dwarfs and about 12,000 K for DA (H-rich) white dwarfs.

Gilles also found that convection is strongly suppressed once the base of the convection zone reaches the degeneracy boundary. Deeper into the star, electron conduction becomes efficient, flattening the temperature gradient, and convection ceases just below the degeneracy boundary. The base of the convection zone then retreats toward the surface, tracking the degeneracy boundary, as the star continues to cool. Figure 12.5, from Fontaine's original 1973 calculation,[11] shows how the amount of mass in the convection zone varies as the luminosity decreases for a 0.612 M_{Sun} white dwarf with a helium envelope. The He convection zone in such a non-DA white dwarf reaches the degeneracy boundary—and therefore its maximum depth—by the time the effective temperature has fallen to about 12,000 K; similarly, the H convection zone in a DA reaches the degeneracy boundary and its maximum depth near 5,200 K.

In either case, when the convection zone reaches the degeneracy boundary—an event termed "convective coupling"—there is a temporary slowing of the white dwarf cooling rate as the star begins to lose thermal energy from the core more efficiently. Similar results were obtained in 1979 by Italian stellar-evolution theorists Francesca D'Antona and her husband Italo Mazzitelli using the Magni-Mazzitelli equation of state.[12]

Gilles' calculations also showed that the relation between the luminosity and core temperature is modified only for models cool enough for the inner boundary of the convection zone to reach the degenerate core. In contrast to the assumption in Mestel's theory of white dwarf cooling, however, the temperature at the degeneracy boundary proved *not* to be a good estimate of the central temperature

[11] Data from Fontaine, G., and Van Horn, H. M. 1976, *Astrophys. J. Suppl.*, **31**, 467, "Convective White-Dwarf Envelope Model Grids for H-, He-, and C-Rich compositions." Plotted by the author.

[12] D'Antona, F., and Mazzitelli, I. 1975, *Astron. & Astrophys.*, **42**, 165, "White Dwarfs External Layers. I. Convection and Central Temperature."

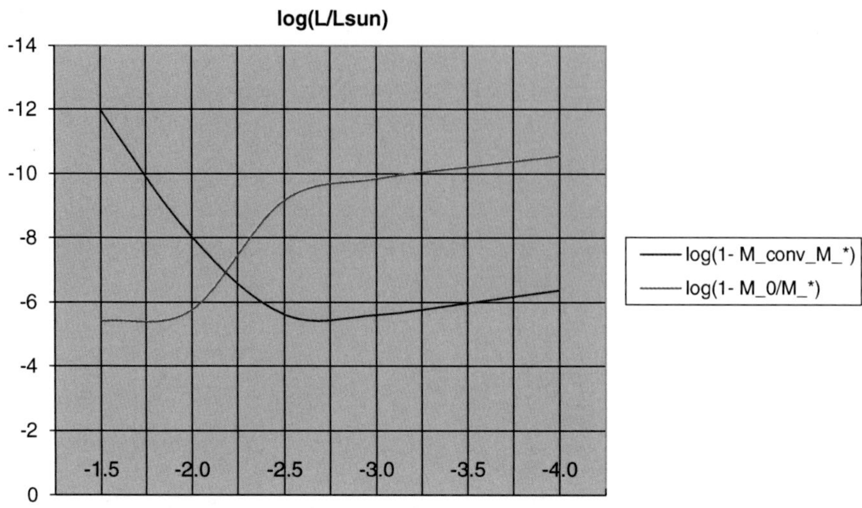

FIGURE 12.5 Variation of the amount of mass in the convection zone of a 0.612 solar mass white dwarf with a pure helium surface layer as the luminosity of the star decreases. The vertical axis is $\log(1-M_r/M_*)$, where M_r is the mass contained within a radius r and M_* is the stellar mass. The zero of this scale is the center of the star, while the top of this axis is within about $10^{-14} M_{Sun}$ of the surface. The horizontal axis is the logarithm of the stellar luminosity, ranging from about $0.03 L_{Sun}$ down to $10^{-4} L_{Sun}$. The blue curve shows the lower boundary of the surface convection zone, while the red curve shows the upper boundary of the degenerate core of the star. As the star fades below about $0.03 L_{Sun}$, the convection zone penetrates deeper into the star, reaching a maximum depth of about $10^{-6} M_{Sun}$. When the convection zone penetrates into the degenerate core, however, the increasing degeneracy in the lower part of the convection zone gradually makes electron conduction more efficient than convection in transporting energy, and the lower convective boundary thereafter retreats gradually upward, paralleling the growth of the degenerate core. Data from Fontaine (1973) and Fontaine and Van Horn (1976)

of the (nearly) isothermal core of a white dwarf. While the degeneracy-boundary temperature does indeed drop rapidly after the base of the convection zone crosses the degeneracy boundary, as Böhm and I had originally found, the central temperature of the white dwarf varies much more gradually, as shown in Figure 12.6.[13]

[13] Data from Fontaine and Van Horn 1976, *op. cit.*, plotted by the author.

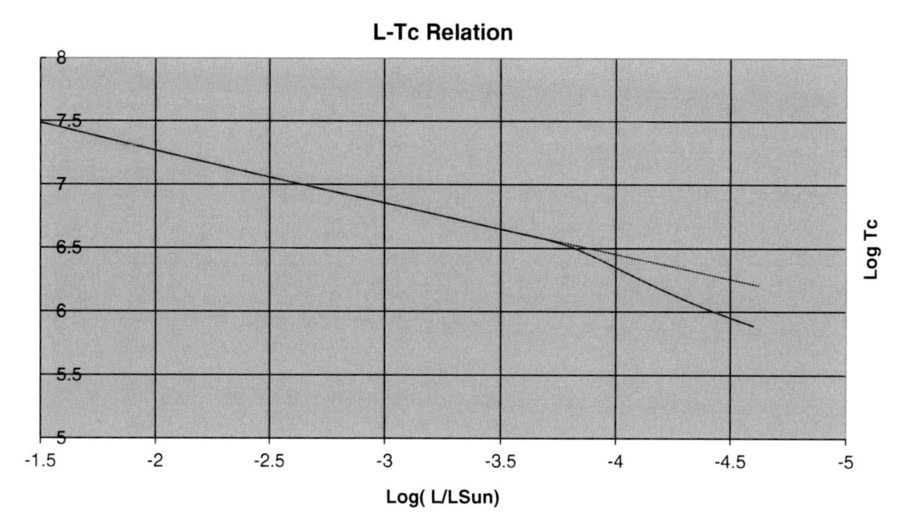

FIGURE 12.6 The logarithm of the central temperature T_c as a function of the logarithm of the luminosity in solar units for a $0.612 M_{Sun}$ white dwarf with a helium envelope. The dotted line is the extension of the trend line from higher luminosities to show the effect of the surface convection zone in depressing the core temperature in cooler white dwarfs. Data from Fontaine and Van Horn (1976)

Even before he had completed his thesis research, Gilles had provided his envelope models to his graduate-student colleague Don Lamb, who used them in his own thesis calculation of white dwarf cooling. After leaving Rochester, Gilles first took a postdoctoral position with John Landstreet at the University of Western Ontario and rapidly became involved with him in observational work on white dwarfs. Gilles' dream, however, had always been to obtain a faculty position at the Université de Montréal, and he jumped at the chance when it was offered to him in 1977. Since then, he has become a world leader in white dwarf research, attracting other top-flight faculty colleagues, producing a steady stream of impressive research in collaboration with colleagues and students alike, and receiving many honors for his work.

13. Leaping into Space

On October 4, 1957, the world awoke to the dawn of the Space Age. Earlier that day, the Soviet Union had for the first time in history launched a manmade artificial satellite into orbit around Earth.[1] At one stroke, Sputnik I moved space travel space from the realm of science fiction into a very present reality. It marked the beginning of humankind's progress from the surface of Earth into space, in a journey that continues to this day. It heralded revolutions in communications, weather forecasting, and an understanding of the universe in which we live.

For astronomers, the advantages of going into space included the ability to carry out observations at ultraviolet and infrared wavelengths, which are blocked by Earth's atmosphere and cannot be studied from the ground, together with the opportunity to get above the atmospheric turbulence, gaining greatly improved angular resolution in images of distant astronomical objects.

The month after the successful orbiting of Sputnik I by the Soviets, which had been shrouded in secrecy, the United States attempted to launch a new rocket called "Vanguard," with full television coverage.[2] The nation was chagrined and embarrassed when the rocket exploded on the launch pad. The first American satellite to reach orbit, *Explorer I*, was delivered into space in January 1958 by a Jupiter C rocket developed by the German rocket scientist Wernher von Braun (1912–1977) and his colleagues, at this point working for the U.S. Army. The instruments aboard this satellite enabled University of Iowa professor James A. Van Allen

[1] Information about Sputnik I can be found in http://history.nasa.gov/sputnik/; accessed 5 August 2012.

[2] The failure of the first Vanguard launch is described in http://en.wikipedia.org/wiki/Project_Vanguard; accessed 19 May 2014. The early history of the U. S. space program is described by John E. Naugle and John M. Logsdon in Logsdon (2001), pp. 1 ff., "Space Science: Origins, Evolution, and Organization."

H.M. Van Horn, *Unlocking the Secrets of White Dwarf Stars*,
Astronomers' Universe, DOI 10.1007/978-3-319-09369-7_13

(1914–2006) to discover Earth's radiation belts, which were subsequently named in his honor.

In July of 1958, President Eisenhower created the National Aeronautics and Space Administration (NASA) to oversee the nation's civilian space program, giving new responsibilities to the former National Advisory Committee for Aeronautics and elevating it to the status of a federal agency.

In 1961, the Soviets shocked the world again by becoming the first nation to send a man into space, when cosmonaut Yuri Gagarin (1934–1968) successfully orbited Earth and returned to a hero's welcome in April of that year.[3] The very next month, President John F. Kennedy made this announcement: "I believe that this nation should commit itself to achieving the goal, before this decade is out, of landing a man on the moon and returning him safely to the earth."[4] The race to the Moon was on, with the United States and the U.S.S.R. vying to become the first to achieve that daunting goal.

NASA now had a dual mission: to continue the exploration of Earth and the rest of the Solar System from space using increasingly sophisticated robotic spacecraft and also to put men—and later, women as well—safely into space.

While both the United States and the Soviets succeeded in sending robotic spacecraft to the Moon—the Soviets becoming the first to image the lunar farside, giving them the right to name the most prominent crater there Tsiolkovsky, in honor of the father of Russian rocketry—the U.S. effort culminated in the *Apollo 11* mission to the Moon, when astronaut Neil Armstrong (1930–2012) stepped down from the Lunar Excursion Module onto the lunar surface on July 20, 1969, and uttered the immortal words, "That's one small step for a man, one giant leap for mankind."[5]

According to Nancy Grace Roman, who oversaw much of the space agency's early astronomy and astrophysics program:[6]

NASA's first orbiting [satellite] missions designed to study the Sun, the Orbiting Solar Observatories (OSOs), were able to provide

[3] http://en.wikipedia.org/wiki/Yuri_Gagarin; accessed 19 May 2014.

[4] http://www.jfklibrary.org/JFK/Historic-Speeches.aspx ; accessed 19 May 2014.

[5] http://en.wikipedia.org/wiki/Neil_Armstrong#First_Moon_walk; accessed 19 May 2014.

[6] The history of space-based astronomy is described by Nancy G. Roman in Logdson (2001), pp. 501 ff, "Exploring the Universe: Space-Based Astronomy and Astrophysics," esp., pp. 514 ff.

reasonable three-axis pointing by locking onto the Sun ... The first satellite to provide versatile three-axis pointing was the first of NASA's major astronomy missions, [the] Orbiting Astronomical Observatory (OAO-1) [launched in 1966] ... On OAO-3, a photomultiplier that measured each point [in a spectrum] individually was scanned across the spectrum ... Intensified vidicons (a space variant of a television camera) were used in several satellites, including OAO- 2 [launched in 1968] and the International Ultraviolet Explorer (IUE) ... An OAO mission launched in 1972 became known as Copernicus .... Until the launch of NASA's Far-Ultraviolet Explorer (FUSE) in 1999, the Princeton spectrometer [aboard Copernicus] was the only free-flying satellite that could observe the far UV ... ."

In 1976, Malcolm Savedoff and I, together with our colleagues François Wesemael (1954–2011), Larry Auer, Ted Snow, and Don York, were able to use the *Copernicus* satellite (see Figure 13.1) to obtain the first ultraviolet spectrum of the white dwarf Sirius B (WD 06422-166) at wavelengths just short of the atmospheric cutoff around 3,100 Å. Savedoff had long been interested in trying to detect UV radiation from this star, and when the *Copernicus* satellite was launched, he proposed a unique observing program to accomplish this.

There was no difficulty in locking onto the Sirius system, since Sirius itself is such a bright star, and to accomplish the

FIGURE 13.1 The *Copernicus* spacecraft. Public domain image, courtesy of NASA

detection of Sirius B, Malcolm proposed a two-part observation. First, with the slit of the spectrograph oriented along the line between Sirius A and Sirius B, he proposed measuring the spectrum of both stars combined. Second, we would measure the spectrum of Sirius A alone at a time when the spectrograph slit was oriented more than 16° away from the A-B axis, so that the spectrum of Sirius B would not be included. Then the spectrum of Sirius B would be obtained by subtracting the digital spectrum of Sirius A from that of A plus B.

After his observing proposal was accepted, Savedoff wrote a program to accomplish these observations and sent it to the *Copernicus* program managers at Princeton. There it was vetted, and in due course the observations were carried out, and the data were transmitted back to Earth. Sometime later, Malcolm and I traveled together to Princeton to carry out the data reductions. We spent a couple of days in a large, open room designated for use by visiting scientists, working at computer terminals to reduce the satellite data and making sure that we had all the information we needed, including the necessary calibration data.

We also carried out preliminary analyses, so that by the time we returned to Rochester we were certain that we had in fact detected Sirius B. We found a clear detection of UV flux coming from the white dwarf in the region around 1,120 Å, as well as a probable detection of the short-wavelength wing of a broad, strong absorption feature at the location of the 1,216 Å Lyman α line of neutral hydrogen. Comparing our measured UV flux with the best model-atmosphere calculations then available, we obtained for Sirius B the effective temperature $T_{\rm eff} = 27{,}000 \pm 6{,}000$ K and a surface gravity given by $\log g = 8.6 \pm 0.2$.[7] This temperature determination was substantially lower than the 32,500 K value obtained in 1971 from high-quality optical observations obtained with the 200-in. telescope at Mount Palomar.[8] The new *Copernicus* data implied a radius of 0.007–0.011 solar radii for

[7] Savedoff, M. P., *et al.* 1976, *Astrophys. J.*, **207**, L45, "The Far-Ultraviolet Spectrum of Sirius B from *Copernicus*."

[8] Greenstein, J. L., Oke, J. B., and Shipman, H. L. 1971, *Astrophys. J.*, **169**, 563, "Effective Temperature, Radius, and Gravitational Redshift of Sirius B."

FIGURE 13.2 Jay Holberg. Recent image courtesy J. B. Holberg. Reproduced with permission

Sirius B, consistent with the zero-temperature white dwarf models of Hamada and Salpeter.

Confirmation of the improved temperature determination for Sirius B became possible in 1983 when Jay Holberg (Figure 13.2)—then serving as a project scientist for the *Voyager 1* and *2* spacecraft—decided to target the Sirius system. As Holberg reports:[9]

> *When I first saw the spectrum … I was initially surprised to see a prominent bump in the flux between 1,000 Å and 1,150 Å, that was not present in the spectrum of Vega [a star that is nearly identical to Sirius in spectral type and which Holberg had observed earlier]. [I realized that the] Voyager instruments were picking up the ultraviolet signature of Sirius B, and that at these short wavelengths the tiny white dwarf was actually outshining the vastly larger Sirius A … . I was intrigued enough to spend some time analyzing the Sirius*

[9] Holberg (2007), p. 193.

FIGURE 13.3 The *International Ultraviolet Explorer (IUE)* spacecraft. Public domain image, courtesy of NASA

> *spectrum and eventually came up with a new temperature for Sirius B of less than 27,000 K. This represented a substantial refinement of the Copernicus temperature estimate of 27,000 ± 6,000 K …*

One of the most productive spacecraft, from the point of view of its impact on astronomers' knowledge of white dwarfs, was the *International Ultraviolet Explorer (IUE)* spacecraft (Figure 13.3).[10] Launched in 1978, this was one of the longest-lived NASA missions, until its shutdown in 1996.

Although it had only a 45-cm-diameter telescope, the *IUE* was equipped with both a low-resolution and a high-resolution spectrograph. Each operated with either short wavelength (approximately 1,150–2,000 Å) or long wavelength (1,900–3,200 Å) cameras. The term "far-UV" generally refers to this interval between

[10] The *IUE* mission is described by Boggess, A., and Wilson, A., in Kondo, *et al.* (1987), p. 3, "The History of IUE." They also include a compendium of scientific results. Thousands of spectra obtained by this versatile spacecraft remain available for use through the *IUE* data archive maintained at NASA's Goddard Spaceflight Center.

the Lyman α line (1,216 Å) and the 3,100 Å limit, where Earth's atmosphere begins to block radiation of shorter wavelengths.

Many hot white dwarfs (and other UV-bright stars called "hot subdwarfs") were found to exhibit rich far-UV spectra, with absorption lines originating from their photospheres or from outflowing winds or circumstellar gas. The photospheric lines could be used to determine the radial velocity, gravitational redshift, rotation rate, and magnetic field of a white dwarf. Lines from ions of such elements as C, N, O, Si, Fe were also found in these spectra.

Often surprising discoveries were made in *IUE* spectra of cooler white dwarfs as well. Many of the DC white dwarfs, featureless at optical wavelengths, were found to reveal neutral atomic carbon (C I) or molecular C_2 features in *IUE* spectra. Thus, these DC white dwarfs were found to be related to those of spectral type DQ—they are stars with He-rich atmospheres that are too cool to show He I lines but display traces of C dredged up to the surface from the degenerate C/O core via deep convective mixing. Some DZ stars also showed rich metallic-line spectra at these wavelengths.

The launch of the *Einstein Observatory (HEAO-2)* on November 13, 1978, and the *EXOSAT* observatory, launched on May 26, 1983, enabled studies of white dwarfs at the still shorter X-ray wavelengths. Two of the very hottest known white dwarfs with H-poor atmospheres were found to be strong X-ray sources: the prototypical DOV pulsating variable star PG 1159-035 (GW Vir = WD 1159-034), with effective temperature greater than 100,000 K,[11] and H1504 + 65, which shows lines of neither H nor He and which is probably hotter than 150,000 K.[12] The surface temperatures of these stars are some 15–25 times hotter than that of the Sun, equivalent to temperatures that are reached only some 15,000–25,000 km deep inside the Sun!

The *Hipparcos* satellite, launched on August 8, 1989, was designed to undertake astrometry from space. Its purpose was to obtain, from repeated observations of the entire sky, the most

[11] Kahn, S. M., *et al.* 1984, *Astrophys. J.*, **278**, 255, "Photospheric Soft X-Ray Emission from Hot DA White Dwarfs;" Petre, R., Shipman, H. L., and Canizares, C. R. 1986, *Astrophys. J.*, **304**, 356, "Evidence for a n(He) / n(H) Versus T_{eff} Coorrelation."

[12] Nousek, J. A., *et al.* 1986, *Astrophys. J.*, **309**, 230, "An Extraordinarily Hot Compact Star Devoid of Hydrogen and Helium."

accurate, consistent, simultaneous solutions possible for the program stars. The mission ended in August 1993. Subsequent data analysis by two independent groups required 3 more years, and resulted in substantial improvements in the astrometric precision in the positions, parallaxes, and proper motions, to about[13] 1 mas or 1 mas year^{-1}, respectively. The final program contained nearly 120,000 stars. Only the brightest few dozen white dwarfs were reached with useful precision, but they included Sirius B. According to Holberg,[14] *Hipparcos* "measured the parallax of Sirius A to be 379.22 ± 1.58 m-arc seconds … . The weighted mean of ground-based and *Hipparcos* parallaxes is 380.02 ± 1.28 m-arc seconds. This corresponds to a distance of 2.8 pc, or 8.4 light years … ."

For the most part, the use of space for astronomy proceeded with little interaction with the manned space program. There were a few important exceptions. Following the end of the Apollo program in 1972, the U.S. effort to maintain a human presence in space focused on the *Skylab* missions, which orbited Earth from 1973 to 1979. Observations by *Skylab* astronauts in 1976 led to the discovery of the very hot DO white dwarf HD 149499B (WD1634-573), which is a companion to a bright K dwarf star.[15] It would be one of the brightest white dwarfs in the sky at visual wavelengths were it not, like Sirius B, hidden in the glare of its cooler companion.

Also in the 1970s, coinciding with a detente in U.S.-Soviet relations, American astronauts and Russian cosmonauts conducted joint exercises in space through the *Apollo-Soyuz* program, employing previously unused hardware left over from the Apollo missions to the Moon, mated in space to the *Soviet Soyuz spacecraft*. Observations made with a telescope deployed during this mission led to the detection at EUV wavelengths of the previously known hot DA white dwarfs HZ43 and Feige 24 (WD 0232+035), each with an effective temperature greater than 50,000 K, and Sirius B near 28,000 K.[16]

---

[13] One milli-arc second (1 mas) equals one thousandth of a second of arc.

[14] Holberg (2007), p. 199.

[15] Parsons, S. G., *et al.* 1976, *Astrophys. J.*, **206**, L71, "*Skylab* Ultraviolet Stellar Spectra: A New White Dwarf: HD149499B."

[16] Margon, B., and Bowyer, S. 1975, *Sky & Tel.*, **50**, 4, "Extreme-Ultraviolet Astronomy from *Apollo-Soyuz*."

FIGURE 13.4 The *Hubble Space Telescope (HST)*. Public domain image, courtesy of NASA

Another chapter in the manned exploration of space began in 1982 with the first operational flight of the Space Shuttle (officially called the "Space Transportation System," or STS). The Shuttle was designed to carry a crew of 5–7 astronauts—both men and women, with the late Stanford-educated physicist Sally Ride (1951–2012) becoming the first American woman in space,[17] riding aboard STS-7 *Challenger* in 1983—together with heavy payloads carried aloft in the Shuttle's roomy cargo bay. In the 30-year period ending in 2011, the STS flew 135 missions.

From the point of view of astronomy, the most important Shuttle mission—which also was the manned mission that had the greatest impact on the field—was the 1990 launch of the *Hubble Space Telescope (HST)* (Figure 13.4) aboard the Shuttle *Discovery* on STS-31. One of NASA's so-called "Great Observatories" missions, the space telescope grew out of a 1946 paper by renowned Princeton astronomer Lyman Spitzer, Jr. (1914–1997).[18]

[17] http://en.wikipedia.org/wiki/Sally_Ride; accessed 19 May 2014.

[18] Spitzer, Lyman 1946, "Astronomical Advantages of an Extra-Terrestrial Observatory," (Project RAND: Santa Monica, CA).

After decades of planning and preparation, the *HST* employed a mirror nearly 8 ft in diameter, the largest ever launched into space. In orbit, its view of the Universe was not hampered by the turbulence or opacity of Earth's atmosphere, giving it a view of astronomical spectra extending from near-infrared wavelengths through the visual portion of the spectrum to near-ultraviolet wavelengths. With such capabilities, it is no wonder that astronomers were eager for the telescope to be deployed from the Shuttle bay, turned on, and begin to return images and spectra from space. Consequently, the entire U.S. astronomical community was appalled when the first images proved to be even blurrier than images obtained from ground-based optical telescopes. What could possibly have gone wrong?

The problem was soon traced to the main telescope mirror itself. Although it had been exactingly ground and polished to an extremely precise shape, that shape turned out to be ever so slightly wrong; it was about two millionths of a meter too flat from the center to the edge.[19] Optical engineers immediately went into overtime to devise a correction: a couple of small mirrors inserted into the optical path—called *COSTAR (Corrective Optics Space Telescope Axial Replacement)*—proved to be sufficient. They were delivered aboard the Shuttle *Endeavour* in mission STS-61 in 1993 and installed by astronauts during a spacewalk. To the relief of the entire astronomical community—not to mention the huge relief of NASA's program managers!—this correction fixed the problem, enabling the *HST* to return stunning images from space and to become one of astronomy's most productive instruments. Ultimately, *HST* was serviced by four Shuttle missions over the decade 1993–2002.

The original spectrophotometric instruments launched with the *HST* were the *Faint Object Spectrograph (FOS)* and the *Goddard High Resolution Spectrograph (GHRS)*, respectively for low- and high-resolution spectrophotometry. They were used extensively in the early years of the *HST*, since the uncorrected optics compromised planned imaging programs more than spectroscopy, before the implementation of the *COSTAR* optics in

[19] http://en.wikipedia.org/wiki/Hubble_Space_Telescope#Flawed_mirror; accessed 19 May 2014.

1993. The first extensive *HST-FOS* low-resolution spectra of white dwarfs were published in 1995. The installation of the newer *Space Telescope Imaging Spectrometer (STIS)* in 1997 allowed spectra of higher quality (and of extended objects) to be obtained.

In 2001, the *Hubble Space Telescope* was used to obtain images of a number of binary systems containing white dwarf companions.[20] In nine of the thirteen systems studied, it was actually possible to obtain images of the white dwarf. In addition, the *Hubble Space Telescope* has provided the very best determination of the radius of Sirius B to date. According to Holberg, it is "0.0084 ± 0.0001 [solar radii,] ... about 5840 km, or 92 % the radius of the earth."[21]

Holberg also gives a fascinating first-hand account of the use of the *Hubble Space Telescope* to determine the gravitational redshift of Sirius B.[22] As he tells the story:

> *The observations ... had to wait for a number of things to fall into place ... . In 1999, during the third Hubble-servicing mission, the original high-resolution spectrometer was replaced with a more capable instrument, the Space Telescope Imaging Spectrograph, or STIS... [In 2003, a] group of astronomers led by Professor Martin Barstow of the University of Leicester in the United Kingdom, and including myself and Dr. Howard Bond of the Space Telescope Science Institute and others, decided the time was ripe to propose to use STIS to make a precise measurement of the gravitational redshift of Sirius B ... . Once a proposal is accepted, a process of detailed planning begins. On the observer's side this means specifying such things as the precise pointing of the telescope, the exposure times, and grating settings, etc. For those in charge of the spacecraft, these inputs are converted into an exact set of commands to point the telescope, acquire the target, and direct the instrument to take and record the observations. Observations are all placed in queue and are executed automatically at the appointed time ... . Our observations began at 1:56 p.m. Greenwich Mean Time on February 6, 2004 ... . while still on the daylight side of the earth, Hubble slewed to a point near earth's limb where Sirius would soon appear. At the same time STIS was configured to per-*

[20] Holberg (2007), p. 121.

[21] Ibid., p. 203.

[22] Ibid., pp. 208 ff.

form our observations. As the spacecraft was entering the earth's shadow, somewhere over the Indian Ocean, its Fine Guidance System acquired a preselected set of guide stars and made fine pointing adjustments to place Sirius B in the large aperture of the STIS spectrometer... These procedures required about 8 minutes out of the approximately 50 minutes of nighttime observing available during each orbit. Our first exposure... provided a spectrum that extended from 3,900 Å to 5,000 Å. This gave good spectral coverage of most of the star's Balmer lines... We would use these data to determine the brightness, temperature, and gravity of Sirius B... STIS then changed... to center the spectrum on the broad hydrogen Balmer alpha line of Sirius B at 6,564 Å... This .. [was] the observation that would determine the gravitational redshift. After a 90-second exposure it was all over .... .

Ten days after we had observed Sirius B, I was visiting Martin Barstow... at the University of Leicester. Late one afternoon in his office,... he asked if I would like to see the Sirius B spectra... He had electronically downloaded them several days earlier after they had been reduced, verified, and archived by the Space Telescope Science Institute in Baltimore. As I looked over his shoulder, he displayed the first set of low-resolution spectra containing the higher order Balmer lines. They looked almost exactly like the theoretical models we had used to plan the observations... A quick fit of the model fluxes to the data would tell us exactly how good they were. Martin ran the fits and overplotted the best fitting model on top of the data. The fit was nearly perfect... the temperature and gravity for the best-fitting model was very close to the results we had obtained in 1998 with EUVE [the Extreme Ultraviolet Explorer spacecraft] .... No one had ever obtained data of this quality on any white dwarf, let alone Sirius B .... A few weeks after my visit, Martin sent an email with the result of the gravitational redshift.

However, the value was ten times larger than Holberg had anticipated. "Something was wrong .... ." Holberg continued. "The answer came about a month later in another email from Martin. [Barstow, Holberg, and their colleagues had mistakenly used the wrong wavelength-calibration data for this observation.] Once the models and the observational data were placed on the same scale the results changed dramatically, in the right direction. The final answer was 80.42 ± 4.82 km/s." This was in excellent agreement with the expected value of 75.6 km/s for the Einstein redshift of a white dwarf with one solar mass and a radius 0.0084 times the solar radius, parameters appropriate for Sirius B.

The advent of wholesale faint-object photometry using CCDs, especially from the *Hubble Space Telescope*, has also resulted in the first discoveries of sequences of white dwarfs in globular star clusters (to which we shall return in Chap. 21). Only in the nearest several clusters is it possible to reach several magnitudes down the sequence. Altogether, *HST* observations by several groups have now detected white dwarf sequences in five globular star clusters: M4, Ω Centauri, NGC 6397, NGC 6752, and 47 Tucanae.

In the early 1990s, two satellite observatories performed all-sky surveys at EUV wavelengths (between 100 and 700 Å).[23] The first was the German Roentgen satellite (*ROSAT*), launched in 1990, which enabled astronomers to detect and analyze hundreds of white dwarfs in the EUV and at soft X-ray wavelengths. The second was NASA's *Extreme Ultraviolet Explorer* (*EUVE*). According to Holberg,[24] "It was only [Stuart] Bowyer's dogged determination that succeeded in the 1992 launch of the *Extreme Ultraviolet Explorer (EUVE)* satellite. *EUVE* was a NASA mission to study the sky in three extreme ultraviolet (EUV) wavelength bands between 80 and 760 Å. The satellite's prime mission was to photometrically map the entire sky at these wavelengths, subsequently uncovering 700 sources of EUV radiation. Among these sources was Sirius, which was detected by both the ROSAT Wide Field Camera and by EUVE … ." The Second *EUVE* Source Catalog[25] includes 104 white dwarf identifications of which over 40 were new. Both the *ROSAT* and *EUVE* surveys also found dozens of previously unknown, hot DA white dwarfs that turned out to be previously unknown companions to nearby, bright Main Sequence stars.

The 1999 launch of the *Far Ultraviolet Spectroscopic Explorer (FUSE)* provided a multiyear mission for observations in the nar-

---

[23] As Holberg (2007), p. 120, points out, EUV radiation "cannot penetrate the earth's atmosphere and even has difficulty traveling very far in our own Galaxy before being absorbed by interstellar hydrogen. Nevertheless, hot white dwarfs such as Sirius B radiate fiercely in the EUV…"

[24] Ibid., pp. 204–205.

[25] Bowyer, S., *et al.* 1996, *Astrophys. J. Suppl.*, **102**, 129, "The Second *Extreme-Ultraviolet Explorer*" Source Catalog.

row wavelength band between 900 and 1,180 Å, called the far ultraviolet. A primary purpose of this observatory was to measure deuterium abundances in various lines of sight, with white dwarfs serving as important background sources. As by-products, however, new information was obtained about hot white dwarfs as well. Then in 2001, Jay Holberg and his collaborators utilized this spacecraft to get the first good look at Sirius B in the far UV since Holberg's *Voyager* observations nearly two decades earlier. As Holberg tells the story,[26]

> *[W]e rolled the spacecraft so that the 10×30 arc second rectangular aperture of the spectrometers was centered on Sirius B [which at that time was 5.4 arc seconds away from Sirius A in its orbit] and oriented perpendicular to the line joining the two stars... Sirius B was successfully observed by FUSE for several hours on 25 November 2001. The observations revealed the spectrum of the white dwarf stretching from 900 Å to 1,100 Å and indicated, to our great delight, that the data were virtually free from scattered light from Sirius A.*

Their observations of the far UV radiation flux from Sirius B and of the Lyman lines of hydrogen that fall in this wavelength band provided accurate measurements of the gravitational redshift, surface gravity, and luminosity and yielded improved determinations of the mass and radius for this key white dwarf.

[26] Holberg (2007), pp. 207–208.

14. Decoding the Spectra of White Dwarfs

The spectrum of any star depends upon the physical properties of the atmosphere from which the radiation is emitted, including the stellar effective temperature, surface gravity, and chemical composition. For a given star, astronomers can decode the spectrum to determine the values of these physical quantities by matching the observations against the spectrum predicted by a detailed numerical model. The development of suitable models for the atmospheres of white dwarf stars, however, proved to be a challenging undertaking and required the abilities of many top astrophysicists over several decades.

The noted German astrophysicist Volker Weidemann (1924–2012; see Figure 14.1) undertook the first serious effort to employ an understanding of stellar atmospheres to extract information about the physical properties of white dwarfs from the best-studied spectra. Born in Kiel in northern Germany, near the base of the Danish peninsula, he obtained his doctorate at Kiel University under the direction of Albrecht Unsöld, who was then the director of the university's Institute for Theoretical Physics and a world leader in stellar atmosphere research.

In the late 1950s, Weidemann traveled from Kiel to Caltech, where he spent a sabbatical year working to extract values for the physical properties of white dwarfs from Greenstein's treasure trove of observational data. Weidemann began his work with a careful analysis of the atmosphere of the white dwarf called van Maanen 2.[1] As the photometric colors of this star are similar to those of the Sun, he assumed that the effective temperatures would be similar as well. Then, from the known absolute magni-

[1] Weidemann, V. 1960, *Astrophys. J.*, **131**, 638, "The Atmosphere of the White Dwarf van Maanen 2."

H.M. Van Horn, *Unlocking the Secrets of White Dwarf Stars*, Astronomers' Universe, DOI 10.1007/978-3-319-09369-7_14

FIGURE 14.1 Volker Weidemann at a 1980 meeting in Erice, Italy. Photograph by the late Robert T. Rood

tude and the assumed effective temperature, he obtained the stellar radius, and from the mass-radius relation for zero temperature white dwarf models he determined the stellar mass and surface gravity. With these basic data, he employed simplified models for the atmosphere to compute profiles for the absorption lines of Fe, Ca, and Mg, and he matched the calculated line profiles against the observations. In this way, he was able to derive realistic values for the physical properties of the star and show that helium is the most abundant element in its atmosphere.

Weidemann next turned to the atmospheres of the DA (hydrogen-rich) white dwarfs.[2] He used the classical DA white dwarf 40 Eridani B, for which the distance, mass, and some other physical properties were well-determined, to establish the temperature scale for these stars. Again constructing simplified models for the white dwarf atmospheres, he computed profiles for the Hγ line of hydrogen, which he compared with observed profiles for the Balmer line obtained by Jesse Greenstein. His results yielded values for the effective temperatures, surface gravities,

[2] Weidemann, V. 1963, *Zeitschrift für Astrophys.*, **57**, 87, "Effektive Temperatur und Schwerebeschleunigung der Weißen Zwerge. Mit 21 Textabbildungen."

stellar radii, and masses for 22 DA white dwarfs. In sharp contrast both to ordinary Main Sequence stars and to van Maanen 2, Weidemann found that the atmospheres of these stars consist of almost pure hydrogen.

At the time of Weidemann's visit to Caltech, digital computers had barely begun to make their appearances in university settings, and computer codes were just beginning to be developed to solve complicated physical problems such as the atmospheres of stars. But these stellar atmosphere codes had not yet begun to be applied to white dwarfs. This situation changed dramatically over the decade of the 1960s, however, when the construction of accurate numerical models for the atmospheres of the stars—including the dependence of the physical quantities on the frequency of the radiation—first became possible because of the advent of powerful digital computers.

In the mid-to-late 1960s, Russell Kurucz at Harvard, and independently Dimitri Mihalas (1923–2013) at the University of Chicago's Yerkes Observatory and Laurence H. Auer at Yale University Observatory, were among the first to develop computer programs to solve these complicated problems and compute the emergent radiation spectra for Main Sequence stars.[3] Kurucz concentrated on incorporating into his computer program—named ATLAS—the enormous amounts of information needed to calculate the absorption and emission of radiation. A typical model atmosphere calculation might require several hundred different frequency points to span the spectrum, as well as the frequency-dependent absorption coefficients for perhaps millions of radiative transitions for numerous different energy levels of perhaps hundreds of different atoms, ions, and molecules. The calculations also require determining the principal stages of ionization for scores of different chemical elements as well as finding which internal energy states of a given atom or ion are populated at each temperature and density.

Auer and Mihalas, on the other hand, focused on creating a computer program to perform automatically all the steps needed in the calculation of a model atmosphere. The method they developed was a self-correcting program capable of bringing a

[3]Kurucz, R. L. 1969, *Astrophys. J.*, **156**, 235, "A Matrix Method for Calculating the Source Function, Mean Intensity, and Flux in a Model Atmosphere;" Auer, L. H., and Mihalas, D. 1969, *Astrophys. J.*, **158**, 641, "Non-LTE Model Atmospheres. III. A Complete Linearization Method."

model to an arbitrarily high degree of precision, in much the same way that computer programs for following stellar evolution were designed to converge a stellar model to arbitrarily high precision. Both their program and ATLAS were designed to calculate the temperature and pressure distributions with depth in a model atmosphere, together with the continuous spectrum emitted at the stellar surface. The narrow absorption lines were computed separately after the fact from the pressure and temperature distributions that had been determined.

In the late 1960s, Satoshi Matsushima (1923–1992) and his colleague Yoichi Terashita applied such methods to compute some of the first digital models for DA (hydrogen-atmosphere) white dwarfs.[4] In their first attempt, they simply employed programs developed to compute stellar atmospheres for Main Sequence stars. Using the resulting models to analyze the observed spectra of white dwarfs, however, led to results for the physical properties of these stars that were strongly at variance with those determined by other methods. The reasons for this disagreement proved to trace back to the very high densities in the surface layers of white dwarfs. As we have already seen, these high densities cause the absorption lines in a white dwarf atmosphere to become extremely broad. In consequence, it is not possible to compute an accurate model atmosphere for such a star without including the line absorption in the calculation. The very broad lines produce absorption over a wide range of wavelengths on either side of the line center, an effect termed "line blocking." Because the radiation that cannot escape through the region of wavelengths affected by the line absorption must escape at other wavelengths—in between the absorption lines—the level of the stellar continuum emission increases slightly at those wavelengths, a phenomenon termed "back warming."

A few years later, Terashita and Matsushima endeavored to take account of the line blocking and back warming produced by the strong Balmer lines in the atmosphere of a DA white dwarf.[5] However, their results still did not agree with the observational

[4] Terashita, Y., and Matsushima, S. 1966, *Astrophys. J. Suppl.*, **13**, 461, "The Structure of White Dwarf Atmospheres. I. Model Atmospheres and Hydrogen-Line Profiles for DA-Type Stars."

[5] Terashita, Y., and Matsushima, S. 1969, *Astrophys. J.*, **156**, 203, "The Structure of White Dwarf Atmospheres. II. Masses and Radii Determined from Atmospheric Parameters of DA Stars."

data. The reason for the remaining disagreement proved to be yet another effect of the high densities in white dwarf atmospheres. In contrast to the case at the much lower densities in the atmospheres of Main Sequence stars, the internal quantum energy levels of the atoms and ions in a white dwarf are strongly affected by interactions with the electrically charged ions and electrons in the atmosphere. These interactions actually lower the ionization potential of the atoms and ions in the surface layers of a white dwarf, and this in turn affects the populations of particles in the different quantum states.

The next step in the development of accurate computer models for white dwarf atmospheres was reported at the conference on white dwarfs that Willem Luyten organized in St. Andrews, Scotland, in 1970. At this meeting, a Caltech Ph.D. student named Harry Shipman (see Figure 14.2) described how he had adapted Kurucz's ATLAS program to compute detailed numerical models for the atmospheres of the DA white dwarfs.[6]

Shipman took into account the pressure ionization of the hydrogen atoms in these dense stellar atmospheres, which Terashita and Matsushima had neglected. This effect turned out to produce appreciably increased line strengths for the Balmer absorption lines in Shipman's models for the atmospheres of the DA stars. The result was that analyses of the observed spectra based on his models yielded values for the surface gravities that were in much better agreement with the theoretical mass-radius relation for white dwarfs than were Matsushima's and Terashita's results.

For the DA white dwarf 40 Eridani B, for example, Shipman fitted the model atmosphere continua he had calculated to scans of the spectral energy distribution that his mentor, the superb instrument designer and astronomer J. Beverley Oke (1928–2004), had obtained with his new 32-channel spectrometer at the prime focus of the 200-in. telescope. The effective temperature they determined in this way was 17,000 K.[7] From the known parallax and apparent magnitude, this led to a stellar radius of 0.0127 ± 0.0006 in units of the radius of the Sun. The resulting

[6] Oke, J. B., and Shipman, H. L., in Luyten (1971), p. 67, "Effective Temperatures of White Dwarfs."

[7] *Ibid.*

FIGURE 14.2 Harry Shipman receiving an award at the University of Delaware in 2008. © University of Delaware. Reproduced with permission

value of the surface gravity—obtained by comparing Shipman's calculations to the observed width of the Hβ line—was $\log g = 7.85$, and from the surface gravity and radius they found the mass to be 0.42 M_{Sun}. For comparison, the mass of this star obtained from the binary orbit a decade and a half earlier was $0.43 \pm 0.04\ M_{\mathrm{Sun}}$.[8]

At the 1970 conference in St. Andrews, it was a young man with fiery red hair and beard who strode confidently to the front of the lecture room to present his findings. Shipman's work left no doubt that model atmosphere calculations for the DA white dwarfs had finally come of age. After finishing his doctorate, Shipman accepted a faculty position at the University of Delaware, where he became one of the leading U.S. astrophysicists to employ model atmospheres to advance our understanding of white dwarf stars

[8] Popper, D. M. 1954, *Astrophys. J.*, **120**, 346, "Red Shift in the Spectrum of 40 Eridani B."

and their spectra. He also displayed a passion for teaching that eventually led him to serve for many years as the Education Officer of the American Astronomical Society.

The first detailed model atmospheres for the DB (helium-atmosphere) white dwarfs also were announced at the St. Andrews conference. Dayal T. Wickramasinghe and Peter A. Strittmatter, both then at the Institute of Theoretical Astronomy at Cambridge University in England, included convective energy transport in addition to purely radiative transfer in their model atmospheres for both the DA and DB stars, and they found that convection is indeed important for DAs cooler than about 13,000 K and for DBs cooler than 25,000 K.[9] The young German astrophysicist Irmila Bues, then at Kiel University, also presented the results of her early DB model atmosphere calculations at this conference.[10] A year or so later, Harry Shipman, too, computed model atmospheres for the DB stars,[11] using the same modified ATLAS code he had employed for his DA model calculations, which he had since modified to include convection, and his results compared well with those obtained by Wickramasinghe and Strittmatter.

These early efforts to incorporate convection into models for the atmospheres of white dwarfs were only partly successful, however. The reason is twofold: First, in the cooler white dwarfs, convection extends upward into the optically thin (semi-transparent) regions of the atmosphere. In consequence, a parcel of matter carrying heat upward as part of the convective flux in the atmosphere cannot retain all of its heat content up to the point where it mixes with its surroundings. Instead, once the parcel starts to become transparent, radiative transfer allows heat to leak out of the mass element, thus reducing the efficiency of convective energy transport. Second, with radiative transfer and convection both playing comparable roles in carrying energy through these upper layers of the atmosphere, the calculation becomes extremely sensitive to the actual temperature gradient through these regions.

This complicated problem was solved in 1971, when Thomas C. Grenfell—then one of Karl-Heinz Böhm's students at the University

[9] Strittmatter, P. A., and Wickramasinghe, D. T., in Luyten (1971), p. 116, "The Line Spectra of White Dwarfs."
[10] Bues, I., in Luyten (1971), p. 125, "Model Atmospheres for Hydrogen-Deficient White Dwarfs."
[11] Shipman, H. L. 1972, *Astrophys. J.*, **177**, 723, "Masses and Radii of White Dwarfs."

of Washington (Seattle)—first developed an automatic method for incorporating convection self-consistently into model atmosphere calculations.[12] This was a significant advance in the development of white dwarf model atmospheres, especially for the cooler of these stars. In the publication reporting his new method, Grenfell makes a point of thanking both his advisor and Erika Böhm-Vitense, who herself had actually developed the now-standard form of mixing-length theory a couple of decades earlier. It is a testimony to the lasting value of this work that Grenfell's method has subsequently been adopted in the majority of leading model atmosphere calculations for cool white dwarfs. This work proved especially important for the cool DZ stars, as it allowed reliable determinations of the abundances of the trace elements in the atmospheres of these non-DA stars.

In addition to Shipman's work at the University of Delaware and Böhm's research group in Seattle, two other groups emerged during the 1970s and 1980s to contribute significantly to the development of white dwarf model atmospheres.

One of these efforts was led by the highly capable, prolific, and productive young German astrophysicist Detlev Koester (Figure 14.3). Born in 1941, Koester carried out his Ph.D. research at Kiel University under Volker Weidemann's direction. His 1971 thesis[13] dealt with the surface layers of white dwarfs and their cooling rates, and his early scientific papers—many co-authored with Weidemann—dealt with the evolution of white dwarfs.

Koester rapidly made himself a leading expert in modeling the outer layers and atmospheres of these stars, hardly a surprising focus for someone trained in the school established by the renowned German astrophysicist Albrecht Unsöld, and he proved to be equally at home with theoretical calculations as with observations. In 1979 Koester—together with his colleagues H. Schulz, and Volker Weidemann—published a grid of models for the classical DA white dwarfs.[14]

---

[12] Grenfell, T. C. (1971); see also Grenfell, T. C. (1972); see also *Astron. & Astrophys.*, **20**, 293, "Model Atmospheres of White Dwarfs with Convection."

[13] Koester (1971).

[14] Koester, D., Schulz, H., and Weidemann, V. 1979, *Astron. & Astrophys.*, **76**, 262, "Atmospheric Parameters and Mass Distribution of DA White Dwarfs."

Figure 14.3 Detlev Koester. Image courtesy D. Koester. Reproduced with permission

During the 1980s, Koester advanced in academic rank as far as he could at Kiel; at that time there were no professorships open. Instead, he accepted a faculty appointment at the Louisiana State University in Baton Rouge, Louisiana, where he joined Arlo Landolt and Ganesh Chanmugam (1939–2004) in continuing work to advance our understanding of white dwarf stars. When his mentor, Volker Weidemann, retired as Professor of Astronomy at Kiel in 1992, Koester was naturally chosen to fill the vacancy, becoming the new director of the Institut für Theoretische Physik und Astrophysik at the University of Kiel.

The other group specializing in calculations for white dwarf model atmospheres began in the late 1970s, when François Wesemael (1954–2011; see Figure 14.4) came to the University of Rochester to undertake graduate work in theoretical astrophysics. Born in Viet Nam, François moved with his family first to France and then to Luxemburg. The Wesemaels finally settled in Canada,

FIGURE 14.4 François Wesemael at the Université de Montréal in 2003. Image courtesy G. Fontaine. Reproduced with permission

where François graduated from the Université de Montréal in 1974. That fall, he entered the University of Rochester, and a year or two later he ended up working with myself and Malcolm Savedoff to construct model atmospheres for white dwarfs.

At that time, Savedoff and I were analyzing our *Copernicus* ultraviolet spectra for Sirius B, and François was given the task of adapting a code provided to us by Larry Auer to produce pure hydrogen models with temperatures covering the range we expected to bracket our observational data. A round-faced, curly-haired young man with an unassuming manner, he was also exceptionally capable, and his quiet demeanor concealed a dry, puckish sense of humor. In 1980, we published the extensive grid of pure-hydrogen models François had constructed, spanning the temperature range upward from 20,000 K.[15] At the highest temperatures we considered, radiation pressure overwhelms even the high surface gravity of a white dwarf, driving strong winds into space from the stellar surface.

[15] Wesemael, F., Auer, L. H., Van Horn, H. M., and Savedoff, M. P. 1980, *Astrophys. J. Suppl.*, **43**, 159, "Atmospheres for Hot, High-Gravity Stars. II. Pure Hydrogen Models."

For his Ph.D. thesis, François carried out extensive modifications of his pure-hydrogen code to enable the construction of pure-helium models, employing the latest results for the continuum and line opacities for the different ionization stages of helium.[16] The new grid of pure He models for very hot DB stars that he constructed again extended in effective temperature from 25,000 K up to very high temperatures, where radiation pressure drives stellar winds and stable models in hydrostatic equilibrium cannot be constructed.

Over the decade of the 1980s, both Wesemael and Koester continued to improve their models, and in 1991 comparisons between their DB model calculations and new ATLAS models constructed by Peter Thejll, Harry Shipman, and their collaborators showed agreement to within 0.2 %.[17] Because the ATLAS code, the Kiel code, and the Auer-Mihalas code had independent origins, this very close agreement provided confidence that all three programs accurately represented the atmospheric structures and the emergent spectra of white dwarf stars.

Upon completing his dissertation in 1979, François joined Gilles Fontaine at the Université de Montréal, and with his departure from Rochester the concentration of effort in white dwarf model atmospheres moved to Montréal as well. Together, François and Gilles and their other Montréal colleagues developed an internationally acclaimed research group that has contributed in major ways to the advancement of our understanding of white dwarfs. They involved students at all levels in their research, and several of the Montreal Ph.D.'s have themselves contributed substantially to white dwarf research.

Although François continued his productive contributions to scientific research on white dwarfs throughout his career, in later life he also became deeply interested in the history of science, particularly in the development of astronomy and astrophysics during the nineteenth and twentieth centuries. His untimely death in 2011 deprived the international white dwarf community of one of its important contributors.

---

[16] Wesemael, F. (1979), Ph.D. thesis, University of Rochester, "Atmospheres for Hot, High-Gravity, Pure Helium Stars."

[17] Thejll, P., Vennes, S., and Shipman, H. L. 1991, *Astrophys. J.*, **370**, 355, "A Critical Analysis of the Ultra Violet Temperature Scale of the Helium-Dominated DB and DBV White Dwarfs."

FIGURE 14.5 Pierre Bergeron. Recent image courtesy P. Bergeron. Reproduced with permission

In 1988, Pierre Bergeron (Figure 14.5), then a graduate student working with Wesemael and Fontaine at Montréal, developed one of the most accurate and detailed grids of LTE model atmospheres for the DA white dwarfs constructed to date.[18] His computer program is based on the code originally created by Auer and Mihalas in the late 1960s and includes the treatment of convection developed in 1972 by Grenfell as well as Wesemael's subsequent modifications.

Bergeron's code is designed to construct models containing different relative abundances of H and He and incorporates bound-free and free-free opacities of all the relevant H and He ions as well as opacities due to the Lyman and Balmer lines of H. In addition, it employs a new equation of state and detailed new opacities from the Opacity Project led by the late Michael Seaton (1923–2007) at University College London. Bergeron subsequently won several prizes for the excellence of his research. At the time of this writing, he is director of the Centre de Recherche en Astrophysique du Québec (CRAQ) at the Université de Montréal.

---

[18] Bergeron, P. (1988), Ph.D. thesis, Université de Montréal, "Propriétés Atmosphériques des Étoiles Naines Blanches Froides de Type DA."

In 1995, recognizing the need for a new generation of models for DB white dwarfs, Alain Beauchamp, another student at the Université de Montréal, constructed a new grid of pure-He model atmospheres for these stars.[19] The code he used was again based on the Auer-Mihalas/Grenfell/Wesemael/Bergeron program. For Beauchamp's pure-He models, the continuum opacity sources include bound-free absorption by all the relevant He ions; absorption by He_2^+ in particular turns out to be important in the temperature range between 6,500 and 16,000 K. Beauchamp also included Thomson scattering by free electrons and Rayleigh scattering by He I and He II. Perhaps the most striking conclusion from his work is the utter lack of gravity discrimination provided by either broad- or narrow-band colors for the DB white dwarfs throughout most of the temperature range they occupy.

As a result of the extensive work by these international teams of scientists, by the last decade of the twentieth century astronomers had acquired the tools needed to extract information about the physical properties of most white dwarfs directly from their spectra. Now we need to ask, what previously hidden physical information have these tools been able to reveal?

---

[19] Beauchamp, A. (1995), Ph.D. thesis, Université de Montréal, "Détermination des Paramétrés Atmosphériques des Étoiles Naines Blanches de Type DB."

15. The Secrets in the Spectra

The rapid progress in stellar atmosphere theory sketched in the previous chapter, coupled with the growth in both the quantity and quality of observational data, finally made possible the efficient decoding of white dwarf spectra. This enabled astronomers to extract increasingly accurate values for the physical properties of white dwarfs. Let us now see what previously hidden "secrets" these tools have revealed.

Effective Temperatures and Surface Gravities

A star's effective temperature determines the shape of its continuous spectrum. Roughly speaking, the hotter the star, the shorter the wavelength at which the spectrum peaks. The surface gravity governs the widths of the absorption lines in the stellar atmosphere. Again, roughly speaking, the higher the surface gravity, the broader are the lines. By careful measurements of the continuous spectrum and the shapes of the absorption lines, an astronomer can thus obtain a star's surface gravity and temperature.

Beginning about the mid-1980s, astronomers became able to obtain very high-quality digital spectra for white dwarfs using CCD detectors. Along with Caltech's Jesse Greenstein, the University of Arizona's James W. Liebert (Figure 15.1) was one of the world leaders in this effort. Born in 1946, Liebert grew up in Coffeyville, Kansas. Attending the University of Kansas as an undergraduate, he became a lifelong, enthusiastic Jayhawks basketball fan. In 1968, he began graduate work at the University of California at Berkeley, but his graduate career was interrupted by the Vietnam War.

Liebert enlisted in the U.S. Navy and spent his military service working in naval intelligence in Washington, D.C. Returning

H.M. Van Horn, *Unlocking the Secrets of White Dwarf Stars*,
Astronomers' Universe, DOI 10.1007/978-3-319-09369-7_15

FIGURE 15.1 James Liebert at a 2006 meeting in Leicester, UK. Photograph by Stefan Jordan. Reproduced with permission

to Berkeley, he completed his doctoral research working with Professor Hyron Spinrad in 1976; his thesis was entitled "Spectrophotometric Studies of White Dwarf Stars." He then accepted a position at the University of Arizona, where he spent the remainder of his scientific career.

A gifted, prolific astronomer, Liebert worked diligently to acquire lists of white dwarf candidates from Willem J. Luyten and then obtained high-quality spectra for large numbers of them. A frequent and valued collaborator with other scientists around the globe, Liebert has played a key role in advancing our understanding of the physical nature of white dwarfs. His insistence on the fundamental importance of observational data, together with his exceptional physical insight, enabled him to design key observational tests of theoretical models for a broad range of astronomical objects. Among other things, he has made major contributions to our understanding of pulsating white dwarfs, magnetic white dwarfs, and the white dwarf luminosity function.

In 1992, together with his colleagues Pierre Bergeron and Rex Saffer, Liebert described a method for determining the effective temperatures and surface gravities for classical DA white dwarfs with high accuracy by fitting theoretical H-line profiles to all of the observed lines simultaneously.[1] The quality of the fits proved to be so good that it is difficult to distinguish between the observational data and the theoretical curves by eye. For the 129 DA white dwarfs they studied, for example, these three astronomers estimated the mean error of their fitting procedure to be 350 K in temperature—out of some tens of thousands of degrees!—and about 0.05 in log g. These results are so accurate that this technique can even discriminate between the small differences in the mechanical structures of zero-temperature and finite-temperature models for white dwarfs.

Five years later, David Finley, Detlev Koester, and Gibor Basri obtained and analyzed CCD spectra for 164 DA white dwarfs in a similar but independent way, and they estimated their values to be accurate to about 1 % in T_{eff} and to within about 0.02 in log g.[2]

Observations from space have also helped to improve the accuracy of white dwarf temperature determinations. Ultraviolet data obtained with the *International Ultraviolet Explorer (IUE)* and the *Hubble Space Telescope (HST)*, when fitted to the best available model atmospheres, yield values of T_{eff} for the DA white dwarfs that agree very closely with those determined from the H lines alone. In his excellent book about Sirius B, Jay Holberg noted that the most accurate determination of this star's temperature has come from a 1996 space observation with the *Extreme Ultraviolet Explorer (EUVE)* satellite.[3] "Because of the extreme sensitivity of the emitted spectrum to temperature," he writes, "it was possible to determine, with great precision, the effective temperature of Sirius B: 24,790 ± 100 K."

Nor have the improvements been confined to the DA white dwarfs. In 1995, Alain Beauchamp used his then-new He-dominated model atmospheres to determine the effective temperatures, surface gravities, and hydrogen-to-helium ratios from observed spectra for 35 individual DB and DBA white dwarfs.[4] He found good agreement between the temperatures determined from

---

[1] Bergeron, P., Saffer, R., and Liebert, J. 1992, *Astrophys. J.*, **394**, 228, "A Spectroscopic Determination of the Mass Distribution of DA White Dwarfs."

[2] Finley, D. S., Koester, D., and Basri, G. 1997, *Astrophys. J.*, **488**, 375, "The Temperature Scale and Mass Distribution of Hot DA White Dwarfs."

[3] Holberg (2007), p. 205.

[4] Beauchamp (1995).

optical and from UV spectra, and his spectroscopic values of log *g* also agree well with those determined from trigonometric parallaxes.

The bottom line is that astronomers can now obtain very accurate values for the effective temperatures and surface gravities of large numbers of white dwarfs directly from the observed spectra.

Luminosities

To determine the luminosity of a white dwarf—the total power output over all wavelengths that it radiates away into space, another of the "secrets" astronomers would like to know—requires three things: the apparent visual magnitude of the star, a quantity called the bolometric correction, and the star's distance. Photometric observations—such as those obtained by Jesse Greenstein and his colleague Olin Eggen at Mount Palomar in the 1960s—give the apparent visual magnitude and UBV colors directly.

The second quantity needed—the bolometric correction—can be obtained from the UBV colors with the aid of white dwarf model atmospheres. By convolving the wavelength-dependent flux from such a model with the experimentally measured wavelength-dependent transmission function for the V-band filter, an astronomer can compute the V-band flux for the model. Similar calculations for the ultraviolet U-band, blue B-band, and infrared I-band filters produce the corresponding fluxes in these wavelength regions. These fluxes enable the investigator to obtain directly the various U-B, B-V, V-I, *etc.*, colors for the model, which can be compared directly to the corresponding photometric colors from white dwarf observations. (Similar calculations can be done for the narrowband Strömgren UVBY colors.) Then, by adding up the model fluxes over all wavelengths, the astronomer can obtain the total radiative flux. The "bolometric correction" is the difference between the absolute bolometric magnitude and the absolute visual magnitude, which is directly proportional to the difference between the logarithm of the total flux and the logarithm of the V-band flux.

Determining the distance to the star, however, is a much more laborious task. This requires measuring its parallax with great precision—a challenging undertaking that necessitates careful measurements over periods of years.

The difficulty and fundamental importance of determining accurate distances to white dwarfs and other nearby stars led Kaj Aa. Strand (1907–2000), then the Scientific Director of the U.S. Naval Observatory (USNO), to establish a program at the observatory in 1964 to make the painstaking observations necessary to determine stellar parallaxes. A Danish astronomer who had moved to the United States in 1938 after receiving his doctorate from the University of Copenhagen, Strand served in the U.S. Army Air Force during World War II. By the time of the St. Andrews meeting in 1970, the USNO parallax program had already yielded parallaxes for 18 white dwarfs, with an accuracy of ±0.004 arc sec, or ±4 milliarcseconds (mas).[5] To help put this in perspective, a white dwarf located 100 pc away from the Sun has a parallax of 0.010 arc sec, or 10 mas. That is about equal to the angular diameter of a dime in New York City as seen from the distance of Washington, D.C. The USNO program has continued to be productive and ultimately generated parallaxes for more than 900 stars, including many tens of white dwarfs. When the *Hipparcos* satellite was launched in 1989 with the specific goal of measuring stellar parallaxes from space, its observing program included some 20 white dwarfs. The USNO and *Hipparcos* parallax determinations agree well, and accurate trigonometric parallaxes are now known for nearly 300 white dwarfs.

The apparent visual magnitude and the bolometric correction combine to yield the apparent bolometric magnitude. Knowing the distance then gives the absolute bolometric magnitude, which gives directly the logarithm of the stellar luminosity. This enables astronomers to place the observed white dwarfs in the theorists' H-R diagram ($\log[L/L_{Sun}]$ plotted against T_{eff}), where they can be compared directly to the theoretical models.

Alternatively, the same quantities can be used to place theoretical models in the observers' H-R diagram (the absolute visual magnitude plotted against a color index, also known as a color magnitude diagram). Figure 15.2 is an excellent example.[6]

---

[5] Strand, K. Aa., in Luyten (1971), p. 18, "Parallaxes of White Dwarfs."

[6] Dahn, C. C., in Solheim and Meištas (1999), p. 24, "USNO CCD Parallaxes for White Dwarfs: What's Available Now ... And Future Prospects."

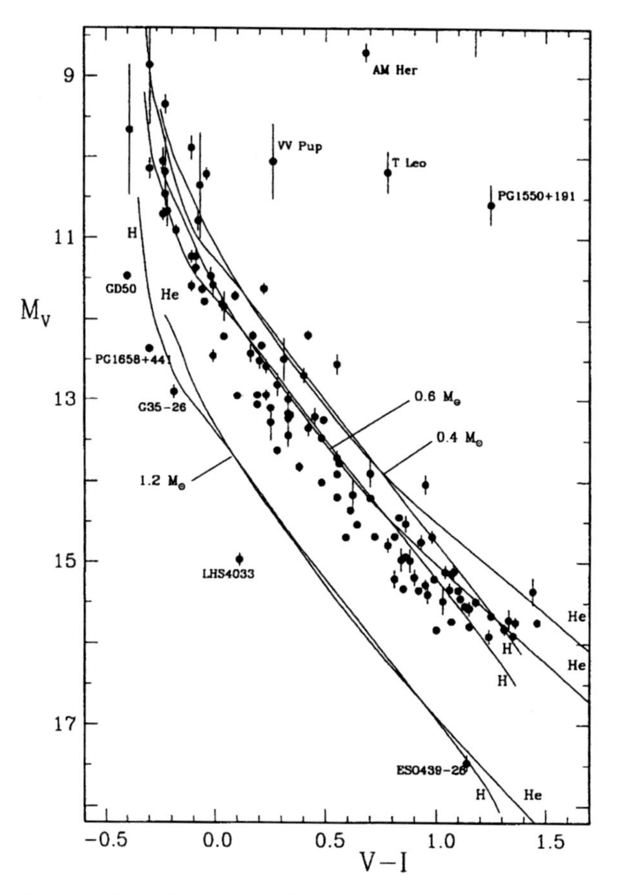

FIGURE 15.2 Comparison between theoretical models and observations in the color-magnitude diagram for white dwarfs from the USNO's parallax program. The vertical axis lists the absolute visual magnitude, with the brightest stars at the top. The horizontal axis is the color index *V–I*, with the bluest stars at the left and the reddest at right. The dots locate positions of individual white dwarfs, some of which are identified, while the curves are white-dwarf cooling tracks computed by Matt Wood for models with three different masses and two different surface compositions (hydrogen or helium). From Dahn 1999. © Astronomical Society of the Pacific. Reproduced with permission

Prepared in 1999 by the U.S. Naval Observatory's Conard Dahn, this figure plots the values of the absolute visual magnitude M_V versus the photometric V-I color for a number of white dwarfs in the USNO parallax program. The vertical bars through some points show typical uncertainties in absolute magnitude resulting from residual uncertainties in the parallax determinations.

A few stars are identified by name. The pairs of curving lines in the figure show theoretical white dwarf cooling tracks for comparison. The uppermost pair of curves corresponds to theoretical models for 0.4 M_{Sun} white dwarfs, the middle pair to 0.6 M_{Sun}, and the lower pair to 1.2 M_{Sun}. One curve in each pair represents models with pure-H atmospheres, as labeled, while the other corresponds to pure-He atmospheres.

As this figure shows, most white dwarfs, no matter whether they have hydrogen-dominated atmospheres (spectral type DA) or helium-dominated atmospheres (the non-DA white dwarfs), cluster near the theoretical curves corresponding to masses $M = 0.6\ M_{Sun}$.

This rather surprising concentration in a narrow range of stellar masses was completely unexpected when it was first discovered. We shall return to this point at the end of the present chapter.

Individual Luminosities, Radii, and Masses and White Dwarf Mass-Radius Relation

With the luminosity and effective temperature of a white dwarf determined as described above, astronomers can obtain the stellar radius directly from the Stefan-Boltzmann law, as they have done for about a century for other types of stars. And if the white dwarf happens to be a member of a binary system—like Sirius B is—its mass can be obtained from the binary orbit. With the mass and radius thus determined independently, astronomers can directly test the theoretical mass-radius relation against the observations.

There are only a few white dwarfs for which the mass and radius can be measured separately, however. Astronomers have endeavored to utilize them in efforts to test the theory ever since Chandrasekhar first derived the theoretical mass-radius relation. As the observational data have improved, the quality of the tests has improved as well. One of the most recent examples was published by the University of Delaware's Judi Provencal and her colleagues in 2002.[7] They used the *Hubble Space Telescope* to obtain

[7] Provencal, J. L., *et al.* 2002, *Astrophys. J.*, **568**, 324, "Procyon B: Outside the Iron Box."

a greatly improved spectrum for the white dwarf Procyon B. Their new data—together with the best determinations for the white dwarfs 40 Eridani B, Sirius B, and Stein 2051 B—are in excellent agreement with the theoretical mass-radius relation, as shown in Figure 15.3.

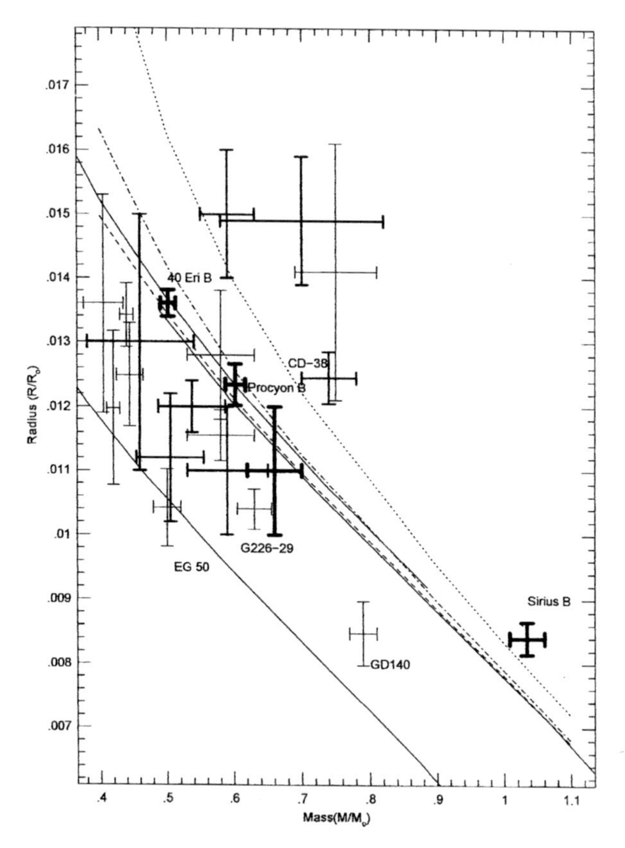

FIGURE 15.3 Test of the white-dwarf mass-radius relation. The crosses represent the uncertainties in stellar radius and mass (both plotted in solar units), with the center of each cross marking the most probable determinations. The heavier crosses represent more certain determinations. In particular, the white dwarfs 40 Eridani B, Procyon B, G226–29, and Sirius B provide the best determinations, and all lie close to the theoretical curves. The solid curves represent Hamada-Salpeter models for compositions of helium, carbon, and iron. The dashed, dot-dashed, and dotted curves show carbon model sequences computed by Matt Wood 1995 with effective temperatures of 7,800 K, 16,700 K, and 25,000 K, respectively. For a fixed temperature, models with lower masses are less degenerate and have proportionately larger radii. From Provencal *et al.* (2002). © AAS. Reproduced with permission

FIGURE 15.4 Judi Provencal. Recent image by Kevin Quinlan, © University of Delaware. Reproduced with permission

Receiving her bachelor's degree from Smith College in 1987, Judi Provencal (Figure 15.4) moved to the University of Texas to pursue her graduate studies, obtaining her Ph.D. in 1994. She subsequently joined Harry Shipman on the astronomy faculty at the University of Delaware, where since 2005 she has been the director of both the Delaware Asteroseismic Research Center and of the "Whole Earth Telescope" (see Chap. 18).

Knowing that real white dwarfs follow the mass-radius relation for theoretical zero-temperature models reasonably closely provides astronomers with another method for determining the individual masses and radii of these stars. Because white dwarfs with lower masses have larger radii, they also have lower surface gravities. And those with larger masses have smaller radii and higher surface gravities. Just as there is a unique, well-defined relationship between mass and radius, there is similarly a unique, well-defined relationship between mass and surface gravity. Using this zero-temperature relation with the spectroscopically determined surface gravity for a white dwarf thus enables an astronomer to estimate unique values for its mass and radius, and spectroscopists were quick to take advantage of this.

Gravitational Redshifts

A completely independent way to determine the mass of a white dwarf is to measure the gravitational redshift of radiation emitted from its surface. This method originated with Einstein's general relativistic theory of gravitation, developed in the early twentieth century. One of the effects this theory predicts is that the wavelength of radiation emitted from the surface of a massive body like Earth or a star becomes gradually longer as the light travels away from the source of the gravitation. Since the gravitational redshift—expressed as an equivalent Doppler shift in velocity—is proportional to the ratio M/R, the stellar mass M can be obtained by combining it either with the spectroscopic radius R or with the surface gravity, which is proportional to M/R^2.

Observationally, the gravitational redshift is the difference between the *measured* velocity of the white dwarf and the *true* velocity along the line of sight (for example, measured from a nondegenerate companion). To obtain sufficient precision for this difficult measurement, an astronomer needs a spectrum with a high signal-to-noise ratio and high spectral resolution, both of which require a fairly bright white dwarf. It is also necessary to correct the measurements for any shifts of the measured line positions due to the high pressures and densities in the white dwarf atmospheres. Fortunately, the pressure shifts of the line cores, which form high in the atmosphere, are very small.

Given the need for high-quality spectra, which can only be obtained with large telescopes, it is not surprising that the first important gravitational redshift work was performed with data from the 200-in. telescope only in the late 1960s. Jesse Greenstein, who then headed Caltech's astronomy graduate program, assigned this as a second-year project to Virginia Trimble (Figure 15.5). Raised in southern California, Trimble was a young woman with a blazing intelligence. Possessed also of a prodigious work ethic, she raced through UCLA in 3 years, graduating in 1964, and then became only the third woman to be admitted to the Caltech graduate program—and the first without an accompanying husband.

Trimble received her doctorate in 1968, taught for a year at Smith College, and joined the faculty at the University of

FIGURE 15.5 Virginia Trimble in 1988. Credit: AIP Emilio Segré Visual Archives, John Irwin Slide Collection

California at Irvine (UCI) in 1971 after a 2-year postdoctoral appointment to Cambridge University in England. In 1972, she married Joseph Weber (1919–2000), who was then pioneering an effort to develop bar detectors for the gravitational waves predicted by Einstein's general relativistic theory of gravitation. Trimble and Weber divided their time between UCI, where she had her primary appointment, and the University of Maryland at College Park, where Weber had his, spending 6 months together at each institution.

Trimble's scientific interests were never confined to white dwarfs, however, and her encyclopedic knowledge of astronomy and astrophysics not only enabled her to contribute to a wide range of topics in astronomy but also made her a superb author of review papers that provided excellent summaries of what was happening in different subfields, a tremendously valuable editor who helped improve many a text, and an award-winning reviewer of scientific articles for many professional journals. The total number of scientific and technical papers she has written is now closing in on 600 and counting, spanning subject areas as disparate as the structure and evolution of stars, galaxies, and the universe; the history of science; Egyptology; Jewish law; and archaeoastronomy.

In their 1967 paper[8] resulting from Trimble's second-year project, Greenstein and Trimble presented the gravitational redshifts they had determined for 92 white dwarfs, and in a subsequent 1972 paper[9] they reported measurements for 74 more. Although this early work suggested a higher mean mass for white dwarfs than that derived from the spectra, the discrepancy was largely resolved in later work. As detection technology continued to advance during the next several decades, a number of other astronomers also made careful measurements of this effect, and by the mid 1990s, they had obtained values for the redshifts of dozens of white dwarfs with accuracies sufficient to obtain useful scientific results.

More recent gravitational redshift measurements for the DA white dwarf 40 Eridani B provide examples of the extent to which the accuracy of this technique has improved over the decades. In 1987, Gary Wegner and I. Neill Reid found the mass of this star to be $M/M_{\mathrm{Sun}} = 0.520 \pm 0.040$ from their gravitational redshift determination.[10] Four years later, Detlev Koester and Volker Weidemann independently measured the gravitational redshift and found $M/M_{\mathrm{Sun}} = 0.554 \pm 0.019$.[11] For comparison, Pierre Bergeron, Jim Liebert, and Mike Fulbright in 1995 determined the mass from their spectroscopic measurements to be $M/M_{\mathrm{Sun}} = 0.516 \pm 0.029$.[12] And in 1998, Judi Provencal and her colleagues used data from the *Hipparcos* astrometric mission to obtain the new value $M/M_{\mathrm{Sun}} = 0.501 \pm 0.011$.[13] It is a tribute to both the excellence of the observers and the quality of the data that these determinations agree to within essentially the limits of the respective error bars.

---

[8] Greenstein, J. L., and Trimble, V. 1967, *Astrophys. J.*, **149**, 283, "The Einstein Redshift in White Dwarfs."

[9] Trimble, V., and Greenstein, J. L. 1972, *Astrophys. J.*, **177**, 441, "The Einstein Redshift in White Dwarfs. III."

[10] Wegner, G., and Reid, I. N., in Philip, Hayes, and Liebert (1987), p. 649, "Radial Velocity Measures of White Dwarfs."

[11] Koester, D., and Weidemann, V. 1991, *Astron. J.*, **102**, 1152, "On the Mass of 40 Eridani B."

[12] Bergeron, P., Liebert, J., and Fulbright, M. 1995, *Astrophys. J.*, **440**, 810, "Masses of DA White Dwarfs with Gravitational Redshift Determinations."

[13] Provencal, J., *et al.* 1998, *Astrophys. J.*, **494**, 759, "Testing the White Dwarf Mass-Radius Relation with *Hipparcos*."

Chemical Compositions

Atmospheric chemical compositions comprise yet another class of "secrets" that astronomers have been able to prise from white dwarf spectra. Spectral classification lifted a corner of the veil shrouding these mysteries, but real progress required sophisticated model atmosphere calculations that first became available during the 1960s. Later in that decade, Karl-Heinz Böhm realized that the surface layers of white dwarfs must possess convection zones that, at least in some cases, must extend up into the atmospheres of these stars. He recognized that convection would affect the temperature gradient through the atmosphere and thus the shapes of the spectral lines in these stars and the conclusions drawn from them. He and his graduate students accordingly set to work to develop methods to include convection in white dwarf model atmosphere calculations. They also undertook careful analyses of the atmospheric properties, including the abundances of the chemical elements in white dwarf surface layers. Prior to their work, the main source of abundance information for white dwarfs had been the analysis carried out by Volker Weidemann before digital calculations for stellar atmospheres became available.

Böhm assigned graduate student Gary Wegner (Figure 15.6) the task of determining the abundances in the white dwarfs using the then-new model atmospheres that included convection. Born in Seattle, Washington, in 1944, Wegner became interested in astronomy as a young boy. Always attracted to observations, he constructed a large telescope in his backyard. After receiving his Bachelor of Science degree from the University of Arizona in 1967, he was immediately accepted into the doctoral program in astronomy at the University of Washington (Seattle).

For his doctoral work under Böhm's supervision, Wegner started with spectra that Böhm had obtained in 1969 for the cool white dwarf van Maanen 2. Wegner also obtained spectroscopic observations of several other white dwarfs using the 84-in. (2.1-m) telescope at the Kitt Peak National Observatory. Three of these stars had strong enough absorption lines to enable him to carry out abundance analyses. For each star, he first computed an accurate model atmosphere, for which the distribution in wavelength of

FIGURE 15.6 Gary Wegner. Recent image courtesy of G. Wegner. Reproduced with permission

the computed emergent flux matched the observed continuum flux distribution. He used this model, together with the best line formation theories then available, to compute accurate profiles to match the observed absorption lines of neutral and ionized calcium (Ca I and II, respectively), as well as Fe I, Mg I, Na I, and Si I.

Comparing the line profiles computed for various assumed abundances of these elements against the profiles actually observed then enabled Wegner to determine the abundances.[14] In this way, he found that the atmospheres of all the stars he studied consist almost entirely of helium. The relative abundance of hydrogen to helium was only about 10^{-4} by number, while the abundances of the metals relative to helium was only about 10^{-9}.

After receiving his doctorate in 1971, Wegner held appointments at several top observatories and universities around the globe before joining the faculty at Dartmouth College in 1982. In the mid-1980s, he became one of a group of astronomers

[14] Wegner (1972).

nicknamed the "Seven Samurai" for their pathbreaking work in identifying an enormous mass of diffuse matter some 250 million light-years away that affects the motions of the Local Group of galaxies near the Milky Way Galaxy, which they dubbed the "Great Attractor." Still at Dartmouth today, Wegner is now a chaired professor of physics and astronomy whose current research interests include galaxies, gravitation, and cosmology.

In the mid-1970s, observations from space in the far ultraviolet and soft X-ray regions of the spectrum, combined with results from the latest model atmosphere calculations pried loose another element-abundance secret that had been hidden in a white dwarf spectrum. In 1974, the *Astronomical Netherlands Satellite (ANS)* detected radiation from Sirius, though the Dutch astronomers could not tell whether it originated from Sirius A or from the white dwarf Sirius B. A few years later, Harry Shipman—by then a professor at the University of Delaware—showed from model calculations that if the atmosphere of Sirius B were to consist solely of hydrogen, with no heavier elements present, it would become increasingly transparent to shorter and shorter wavelengths of radiation, in effect allowing astronomers to "see" into much deeper and hotter layers of the atmosphere at the shorter wavelengths.[15] Shipman also showed that even a few parts per 10,000 of helium or heavier elements would provide sufficient opacity to suppress the short-wavelength radiation. Up to this point in time, astronomers had generally assumed that white dwarf atmospheres contained some proportion of heavier elements, albeit perhaps in amounts too small to be detected directly in the spectrum, in addition to the dominant atmospheric components. Shipman's result at one stroke demonstrated that this is not the case and that, in general, the DA white dwarfs indeed do have nearly pure hydrogen atmospheres.

The following year, the results for Sirius B were joined by similar results for other DA white dwarfs. Because it was detected as a strong EUV source by *Apollo-Soyuz* and by soft X-ray rocket and satellite experiments, HZ 43 was the first white dwarf for which the energy distribution from X-ray to far-ultraviolet

---

[15] Shipman, H. L. 1976, *Astrophys. J.*, **206**, L67, "Sirius B: A Thermal Soft X-Ray Source."

wavelengths was subjected to extensive analyses. Model atmosphere calculations by Larry Auer and Harry Shipman in 1977 provided further support for a pure H composition for the DA white dwarfs. They showed by analysis of the EUV flux from HZ 43 that the He abundance in this white dwarf is much less than the solar value.[16] A decade later this result was confirmed and extended to many more DA white dwarfs by other scientists.

In 1992, Stefane Vennes and Gilles Fontaine added another twist to this story, when they used a grid of stratified H/He model atmospheres to analyze EUV and soft X-ray observations of several hot DA, DAO, and DAB stars.[17] They found that, although all six white dwarfs with effective temperatures less than about 35,000 K have pure H atmospheres, six of eight hotter white dwarfs require He or heavy elements to provide additional opacity. Given the high surface gravities of white dwarfs, however, one would expect element segregation to proceed very rapidly in these stars. So why should these heavier elements be present at all in the surface layers of these stars?

The following year, Leicester University's Martin Barstow and his colleagues confirmed and extended this mystery, when they found only 55 of 384 *ROSAT* sources to be degenerate stars, while they were expecting to detect about 500 white dwarfs with effective temperatures higher than 20,000 K.[18] This apparent shortfall provides additional evidence for opacity in excess of that provided by pure hydrogen alone in the atmospheres of the hotter DA white dwarfs, a finding that was confirmed by the far-UV spectra of stars such as G191-B2B (WD 0501 + 527) and MCT 0455-2812 (WD 0455-282), which contain lines of C, O, N, Si, S, and P in addition to the ultraviolet Lyman lines of hydrogen. Detailed analyses of these spectra indicate that heavy elements are indeed present in the atmospheres of the hot DA white dwarfs, though with abundances that are orders of magnitude smaller than

[16] Auer, L. H., and Shipman, H. L. 1977, *Astrophys. J.*, **211**, L103, "A Self-Consistent Model-Atmosphere Analysis of the EUV White Dwarf HZ 43."

[17] Vennes, S., and Fontaine, G. 1992, *Astrophys. J.*, **401**, 288, "An Interpretation of the Spectral Properties of Hot, Hydrogen-Rich White Dwarfs with Stratified H/He Model Atmospheres."

[18] Barstow, M. A., *et al.* 1993, *Mon. Not. Roy. Astron. Soc.*, **264**, 16, "ROSAT Studies of the Composition and Structure of DA White Dwarf Atmospheres."

they are in the Sun. This puzzle concerning the presence of heavy elements in the atmospheres of hot white dwarfs is taken up in the following chapter.

Mass Distributions

In 1972, John Graham of the Carnegie Institution of Washington became the first astronomer to obtain narrowband uvby photometric colors for white dwarfs.[19] The technique of narrowband photometry—so-called because the filters used in making the photometric measurements are only a few hundred Ångstroms wide, far narrower than the broadband UBV filter bandpasses—had been introduced by Danish astronomer Bengt Strömgren (1908–1987) in 1956 in order to capitalize on improved photodetection technology and better understanding of the properties of stellar atmospheres. What Graham found in 1972 was that, in the two-color u-b *vs.* b-y diagram, the DA white dwarfs followed a locus with a much narrower distribution than did the same stars in the broadband U-B vs. B-V two-color diagram.

The explanation for this finding was provided in a pioneering 1979 paper by Detlev Koester, H. Schulz, and Volker Weidemann.[20] Using Koester's then recently computed DA model atmospheres, they convolved the emergent fluxes for the models with the filter transmission functions for the narrowband uvby filters to obtain theoretical Strömgren color indices. Each model atmosphere is uniquely specified by its effective temperature and surface gravity, and the resulting sets of colors for the models formed an irregular grid in the u-b *vs.* b-y diagram. The grid spacings provided good discrimination in $T_{\rm eff}$ and log *g*. When Koester and his colleagues plotted the positions of the 86 DA white dwarfs they were studying in the same u-b vs. b-y diagram, they found that they all clustered very closely around log g = 7.88 ± 0.33. This was a surprisingly small spread in surface gravity, and—using the theoretical mass-radius relation—it translated into a sharp peak near 0.58 $M_{\rm Sun}$, with three quarters of the stars having masses within 0.15 $M_{\rm Sun}$ of this peak value. A decade and a half later, Bergeron *et al.* derived a new

[19] Graham, J. A. 1972, *Astron. J.*, **77**, 144, "uvby Photometry of White Dwarfs."
[20] Bergeron *et al.* (1992, *op. cit*).

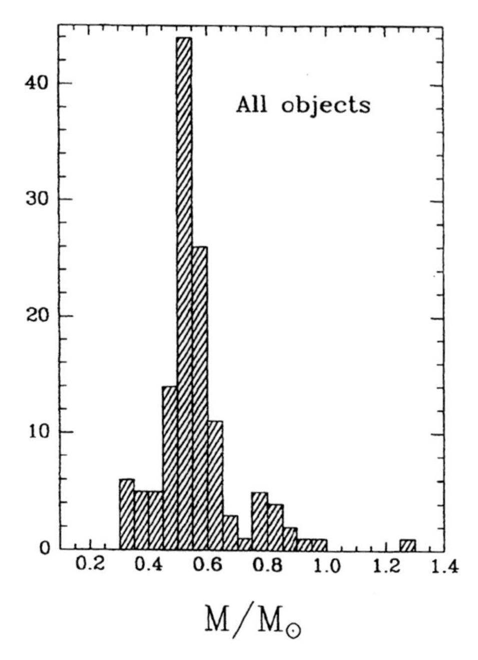

FIGURE 15.7 The mass distribution of the DA white dwarfs, from Bergeron *et al.* (1992). The peak of the mass distribution is 0.56 M_{Sun}, and the low- and high-mass tails of the mass distribution are clearly evident, extending down to about 0.3 M_{Sun} and up to more than 1.2 M_{Sun}. © AAS. Reproduced with permission

generation of mass distributions for 129 DA white dwarfs from the availability of accurate spectroscopic masses.[21] As shown in Figure 15.7, their work confirmed the narrowness of the mass distribution for the DA white dwarfs. Several other groups of scientists have obtained similar results for additional large numbers of DAs. The reasons for this particular peak value and for the sharpness of the mass distribution are among the secrets of the white dwarfs that remain to be decoded.

Analogous work by Alain Beauchamp in 1995, employing the fitting of He I line profiles, showed that the mean mass of the DB white dwarfs is the same, within the errors as that of the DA's.[22] Although the precision of Beauchamp's determination is higher than that of earlier studies using photometric colors, the value he found is quite comparable to earlier results.

[21] Koester, D., Schulz, H., and Weidemann, V. 1979, *Astron. & Astrophys.*, **76**, 262, "Atmospheric Parameters and Mass Distribution of DA White Dwarfs."

[22] Beauchamp (1995, *op. cit*).

16. Understanding the White Dwarf Menagerie

By the late 1950s and early 1960s, the main aspects of white dwarf spectra had become clear, thanks in large measure to the diligent work of Jesse Greenstein and his collaborators using the 5-m telescope on Mount Palomar. As we saw in Chap. 8, about three-quarters of these stars display only the broad lines of hydrogen in their spectra and nothing else at all; they are classified as spectral type DA. The remainder—collectively termed "non-DA" white dwarfs—show no hydrogen at all. Presumably these atmospheres contain mostly helium, the second most abundant element in the Universe. Indeed, the DO and DB stars present only absorption lines due to ionized and neutral helium, respectively. The DQ stars exhibit traces of carbon in their spectra; the DZs show weak lines of metallic elements such as magnesium, calcium, and iron; and the puzzling DC stars display no absorption lines at all! What is going on here? How is one to make sense of all the varied white dwarf spectra, each one utterly different from the spectra of Main Sequence stars?

In 1984, Icko Iben, Jr., proposed a "two-channel" model to account for the two main compositional groups of white dwarfs, suggesting that they originate from the H-rich and H-poor groups of post-AGB stars.[1] Assuming that the end of the "Asymptotic Giant Branch" (AGB) phase of evolution for a particular star occurs when the H-burning shell source is finally extinguished, Iben found that a residual hydrogen envelope with mass of about 10^{-4} M_{Sun} still remains at the surface of the contracting, degenerate star. He argued that such a star will evolve into a DA white dwarf,

[1] Iben, I., Jr. 1984, *Astrophys. J.*, **277**, 333, "On the Frequency of Planetary Nebula Nuclei Powered by Helium-Burning and on the Frequency of White Dwarfs with Hydrogen-Deficient Atmospheres."

H.M. Van Horn, *Unlocking the Secrets of White Dwarf Stars*, Astronomers' Universe, DOI 10.1007/978-3-319-09369-7_16

retaining a hydrogen-rich surface composition through its cooling phase. Alternatively, he pointed out that the remaining hydrogen might be ejected in a final mass-loss event, for example triggered by a helium-shell flash that leaves no hydrogen but retains a residual He envelope with a mass of about 10^{-2} M_{Sun}. Calculations show that about 25 % of He-shell flashes do appear to produce H-deficient central stars of planetary nebulae that subsequently become DO-DB white dwarfs. Indeed, this is roughly the observed percentage of post-AGB stars that are H-poor.

About a dozen years earlier, Peter A. Strittmatter and Dayal T. Wickramasinghe had made another advance in understanding the menagerie of white dwarf spectra. At the time, Swiss astronomer Strittmatter (Figures 16.1) and Sri Lankan astrophysicist Wickramasinghe (Figures 16.2) were at Cambridge University in England. Both subsequently went on to distinguished careers, Strittmatter at the University of Arizona, where he succeeded Bart Bok to serve for decades as the director of Arizona's Steward Observatory and chairman of the Astronomy Department. Strittmatter displayed aggressive, single-minded dedication in building up the facilities of the observatory and in continuing Gerard P. Kuiper's and Bart Bok's drive to build the University of Arizona into a major, world-class center of astronomical excellence. Wickramasinghe, now a distinguished professor at the Australian National University, subsequently also played a leading role in developing our understanding of white dwarfs with strong magnetic fields, as we shall see in Chap. 19.

Strittmatter's and Wickramasinghe's suggestion[2] built upon Karl-Heinz Böhm's then-recent discovery that cool white dwarfs must develop surface convection zones. They pointed out that if the surface H layer were thin enough, the hydrogen could be mixed into the subsurface He convection zone that develops as the star cools, diluting the hydrogen so much that the H lines are no longer visible, thus transforming a DA star into a DB. In this case, *no* DA white dwarfs should be found in the range of effective temperatures where the DBs exist. Although this is not exactly what is observed, Harry Shipman soon pointed out that the basic idea can be salvaged if DA white dwarfs are born with a range of different

---

[2] Strittmatter, P. A., and Wickramasinghe, D. T., in Luyten (1971), p. 116, "The Line Spectra of White Dwarfs."

FIGURE 16.1 Peter A. Strittmatter at a 2009 meeting in Columbus, Ohio. Photograph by the late Robert T. Rood. Reproduced with permission

FIGURE 16.2 Dayal T. Wickramasinghe during a meeting in Tübingen, Germany, in 2010. Photograph by Thomas Rauch. Reproduced with permission

masses for their surface hydrogen layers.[3] Those with sufficiently thin layers could be converted into DBs, while those with thicker H layers would remain DAs.

In fact, astronomers *have* found that the ratio of the DA stars to non-DAs varies with effective temperature, suggestive of just such differences in the mass of the residual hydrogen layers. This was first pointed out by Villanova University's Edward M. Sion and by Caltech's Jesse Greenstein in the mid 1980s.[4] They found that hydrogen-deficient degenerate stars strongly predominate at the highest temperatures, extending up to more than 100,000 K, where the hottest white dwarfs divide into the two basic composition groups, with H- or He-dominated spectra. The number of DAs increases at intermediate temperatures, and *all* white dwarfs in the "DB gap" between about 45,000 and 30,000 K appear to be DA stars.[5] The ratio shifts again in favor of He-atmosphere white dwarfs at the coolest temperatures. The classical DB stars, which show only absorption lines due to neutral He, extend downward in temperature from 30,000 to about 11,000 K. This observational fact—that the DA/non-DA ratio varies with effective temperature—demonstrates conclusively that the spectra of white dwarf stars *do* change as these stars evolve.

Independently, Annie Baglin (Figure 16.3) and Gerard Vauclair (Figure 16.4) in France also pointed out in the early 1970s that convective mixing could transform a DA white dwarf into a DB.[6] They—and Harry Shipman[7]—went even further, showing that at sufficiently low temperatures white dwarfs with He-dominated surface layers become too cool for the absorption lines of neutral He to be visible and would accordingly be classified as DCs. Baglin received her doctorate working on white dwarfs under the supervision of the renowned French astrophysicist Evry Schatzman of the

---

[3] Shipman, H. L. 1972, *Astrophys. J.*, **177**, 723, "Masses and Radii of White Dwarfs."

[4] Sion, E. M. 1984, *Astrophys. J.*, **282**, 612, "Implications of the Absolute Magnitude Distribution Functions of DA and Non-DA White Dwarfs;" Greenstein, J. L. 1986, *Astrophys. J.*, **304**, 334, "The Frequency of Hydrogen White Dwarfs as Observed at High Signal to Noise."

[5] Liebert, J. W., in Hunger and Schönberner (1986), p. 367, "The Origin and Evolution of Helium-Rich White Dwarfs."

[6] Baglin, A., and Vauclair, G. 1973, *Astron. & Astrophys.*, **27**, 307, "Evolutionary Sequence for DA, DB, DC White Dwarfs.

[7] Shipman (1972), *op. cit.*).

FIGURE 16.3 Annie Baglin. Recent image courtesy O. Daudet/Ciel et Espace Photos. Reproduced with permission

FIGURE 16.4 Gerard Vauclair at a meeting in Tübingen, Germany, in 2010. Photograph by Stefan Jordan. Reproduced with permission

Institut d'Astrophysique at the University of Paris. She subsequently took a faculty position at the Observatoire de Nice, where she supervised 16 doctoral students of her own and went on to contribute to several areas of astrophysics.

Baglin is now director of Research Emerita at the CNRS Laboratoire d'Etude Spatiale et d'Instrumentation en Astrophysique (LESIA) and is the principal investigator of the French-led European *CoRoT* space mission to study the oscillations of stars and search for extrasolar Earth-like planets. Her colleague, Gerard Vauclair, also became a major figure in astronomy. Over the years, he has contributed substantially to the understanding of white dwarf atmospheres, the observational determination of element abundances for these stars, new theories and observational data of white dwarf oscillations, and space-based astronomy, among other topics.

Beginning about 1980, astrophysicists came to recognize that several other physical processes, in addition to convection, can affect the distributions of the chemical elements in the surface layers of white dwarfs. At the very highest temperatures, where stars leave the domain populated by the central stars of planetary nebulae in the Hertzsprung-Russell diagram and begin their prolonged cooling, sending them into the realm of the white dwarfs, the flux of radiation from the stellar surface is so high that radiation pressure can even drive winds from the star. This is the last gasp of the post-AGB planetary nebula phase of stellar evolution. Hydrogen and helium are fully ionized at these extremely high temperatures, and heavier elements—such as carbon and metals like Ca, Mg, and Fe—exist in highly ionized states.

The absorption and re-emission of radiation by these species exerts the force that drives these stellar winds. As the temperature declines the radiation fluxes drop, but even when they are no longer high enough to sustain radiation-driven winds, the absorption and re-radiation of photons by the bound states of the heavier elements provides sufficient force to levitate them to higher levels in white dwarf atmospheres. There they can be detected from their characteristic absorption lines. In an important 1979 paper, Gerard and Sylvie Vauclair and Jesse Greenstein at Caltech[8]—and simultaneously but independently Gilles Fontaine and Georges Michaud at the Université de Montréal[9]—first considered this pro-

[8] Vauclair, G., Vauclair, S., and Greenstein, J. L. 1979, *Astron. & Astrophys.*, **80**, 79, "The Chemical Evolution of White Dwarf Atmospheres: Diffusion and Accretion."

[9] Fontaine, G., and Michaud, G. 1979, *Astrophys. J.*, **231**, 826, "Diffusion Time Scales in White Dwarfs."

cess as the source of the trace amounts of heavy elements detected in the hot white dwarfs, and subsequent calculations and observations have generally confirmed their results.

A second process, which acts to deplete heavy elements from white dwarf atmospheres, is gravitational segregation. Evry Schatzmann was again the first to draw attention to this process, in his marvelous small book published in 1958.[10] He pointed out that gravity naturally tends to separate heavier elements, which usually sink, from lighter elements such as hydrogen and helium, which tend to float. The rate at which this diffusive process can occur is limited by collisions among the atoms and ions, which impede the segregation. But the high surface gravities of white dwarfs act to increase the rate, leading to observable changes in the element abundances in white dwarf atmospheres. Accurate computations of the diffusion coefficients required to calculate the rate of gravitational segregation have been carried out by a number of groups, including detailed investigations by Gilles Fontaine's student C. Paquette and collaborators in 1986.[11] They and others showed that the equilibrium state produced when gravitational segregation is complete does *not* contain discontinuities between layers with different compositions but rather contains a "composition transition zone" in which the concentration varies continuously from a nearly pure composition of the lighter element to a nearly pure composition of the heavier element(s) in the deeper levels.

A third process that can affect the observed spectra of white dwarfs is accretion of interstellar matter. Space is filled with enormous clouds of gas and dust. They can appear to observers as wispy reflection nebulae around young stars such as those in the constellation Orion or as dark obscuring regions hiding the light from background stars. The "Fingers of God"[12] in the Eagle nebula, in the famous image from the *Hubble Space Telescope*, are an example

---

[10] Schatzman (1958).

[11] Paquette, C., *et al.* 1986, *Astrophys. J. Suppl.*, **61**, 177, "Diffusion Coefficients for Stellar Plasmas;" *ibid*, p. 197, "Diffusion in White Dwarfs: New Results and Comparative Study."

[12] Also called the "Pillars of Creation."

FIGURE 16.5 The "Fingers of God" in the Eagle nebula, in this famous image from the *Hubble Space Telescope*, are dark obscuring clouds of gas and dust in space. Public domain image, courtesy of NASA

of the latter (see Figure 16.5). As stars move in their orbits around the center of our Milky Way Galaxy, they occasionally encounter such interstellar clouds, and gravity can attract matter to accrete onto a star.

Examples of a perhaps similar process have even been observed within our own Solar System. There are numerous examples of comets being accreted into the Sun. And in 1994, the world watched in fascination as Comet Shoemaker-Levy 9 was accreted onto the planet Jupiter. While most comets reside in a vast, diffuse cloud around the Solar System, they are thought to be remnants of the primordial material from which the Sun and planets formed, and it is probable that solid matter in similar form exists in space between the stars. If similar processes occur as a white dwarf passes through a cloud of gas and dust, accretion of solid particles of matter—composed primarily of heavier elements—can occur, enriching the surface layers of the star in those elements.

Theorists all the way back to Eddington have endeavored to estimate the accretion rates of interstellar matter by stars. Different approaches yield different values for the accretion rates, which may range from perhaps as much as 10^{-15} solar masses per year for accretion from a dense interstellar cloud—equivalent to the accretion of one comet per year about the size of Halley's Comet[13]—down to a rate 100,000 times smaller for accretion from the diffuse interstellar medium.

In 1989, Harry Shipman proposed a modification of Iben's two-channel picture for the evolution of white dwarf spectra that incorporates these physical processes[14] (see Figure 16.6). In this figure, the effective temperature declines from more than 100,000 K at the extreme left-hand side of the figure to less than 5,000 K at the extreme right. This work built on Shipman's suggestion made a decade and a half earlier that stars may leave the AGB with a distribution of residual hydrogen mass at the end of this phase of stellar evolution, rather than having either an amount that is about 10^{-4} M_{Sun} or else zero (the latter implying a pure He surface

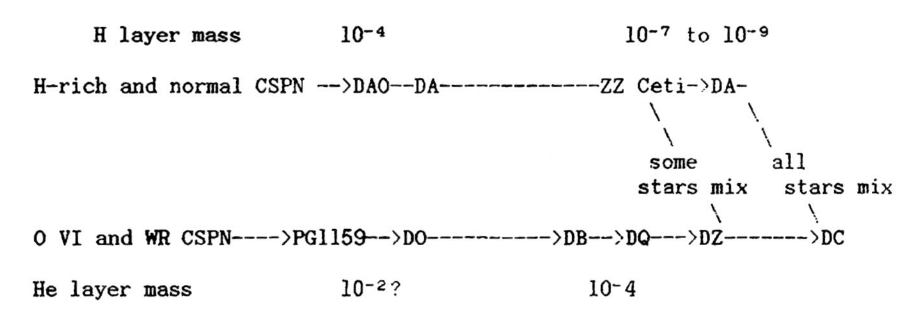

FIGURE 16.6 The modified two channel picture for the formation of white dwarfs with H- and He-dominated spectra, as proposed by Shipman in 1989

[13] The actual process may consist of the accretion of small solid grains or chunks of matter, rather than entire comets.

[14] Shipman, H. L., in Wegner (1989), p. 220, "Origin of DA and Non-DA White Dwarfs."

layer). In all cases, mass loss continues during the high-luminosity phases immediately preceding the white dwarf state.

Let us end this chapter by considering the effects of these several physical processes on the appearance of a post-AGB star as we follow its evolution from the highest temperatures on the cooling track down into the domain of the white dwarfs.

Consider first the spectral evolution of the hottest degenerate stars that begin their final cooling with relatively "thick" surface layers of H (about 10^{-4} M_{Sun}; the left-hand side of Figure 16.6). Radiation-driven mass loss decreases as the effective temperature declines (moving to the right in the figure), but radiative levitation continues to support heavy elements, such as C, N, O, Si, Fe, and Ni, in the stellar atmosphere. Such a star will appear first as a hydrogen-rich central star of a planetary nebula and then as a hot DAO or DAZ white dwarf. Radiative levitation becomes less important as the temperature continues to decrease, and by the time the star has cooled to about 40,000 K, diffusion has caused H to float up to the stellar surface and the heavy elements to sediment out below the photosphere. The white dwarf now appears as a classical hot DA, displaying the Balmer absorption lines of neutral H prominently in its spectrum.

If the mass of the surface H layer is less than about 10^{-15} M_{Sun}[15], it undergoes convective mixing when the effective temperature has declined to about 30,000 K and the He convection zone makes its closest approach to the white dwarf surface. This change defines the "red edge" of the "DB gap," and the star thereafter continues to cool as a classical DB white dwarf. Conversely, if the surface H-layer mass is sufficiently large, the star remains a DA until it becomes cool enough (*i.e.*, the effective temperature falls below about 10,000 K) that other processes must be considered. One such process is again convective mixing, which occurs near this temperature when the surface H convection zone reaches deep enough into the star to mix with the underlying He convection zone (the dashed line labeled "some stars mix" in Figure 16.6).

Just before this transition, a DA white dwarf becomes a pulsationally unstable ZZ Ceti star (see Chap. 17). Because this temperature is too low for He I lines to appear, such a star may

[15] This case is not illustrated in the figure.

continue to appear as a cool DA, even though it may now have a substantial atmospheric He abundance. Another important effect in a cool white dwarf is accretion of interstellar matter. For a star that retains a hydrogen-dominated spectrum, this process converts it into a cool DAZ white dwarf (not shown in the figure). At the lowest temperatures, less than about 5,000 K, the star finally becomes too cool even to display the Balmer lines and is classified as a DC white dwarf (the second dashed line, labeled "all stars mix" in Figure 16.6).

The parallel evolution of the H-deficient degenerate dwarfs begins with the very hot, hydrogen-deficient CSPN, which quickly evolve into objects known as PG1159 stars. These stars have effective temperatures as high as 170,000 K or more, they display prominent spectral features due to CNO elements (and little else), and they are likely to be undergoing continuing mass loss. As happens for white dwarfs with H-dominated spectra, mass loss and radiative support of the heavy elements decline as the star cools and the effective temperature decreases, and diffusion causes the lighter elements to float to the surface. If very little or no hydrogen remains distributed within the He surface layer, the star appears as a hot DO white dwarf (with effective temperature ranging from perhaps 100,000 K down to about 80,000 K). Continued cooling converts hot DOs into cool DOs, which populate the temperature range from about 75,000 K down to about 45,000 K. Any residual H remaining in the star floats to the surface as it cools through this range of effective temperatures (not shown in Figure 16.6).

Accretion of interstellar matter may also add H (and heavier elements) to the star during this period. Accreted heavy elements rapidly settle below the stellar photosphere, while the hydrogen accumulates at the surface. By the time the star's effective temperature has declined to about 45,000 K (just to the left of—and thus at higher temperatures than—the region marked "DB" in Figure 16.6), the combination of diffusive gravitational segregation and accretion have converted it from a DO into a DA white dwarf. This temperature marks the "blue edge" of the DB gap, and the star then continues to cool as a DA white dwarf.

When the star has cooled to $T_{\rm eff}$ at about 30,000 K, a near-surface He convection zone develops. If the residual H-layer mass is sufficiently small (less than about 10^{-15} $M_{\rm Sun}$), the hydrogen is mixed into and diluted by the He convection zone. The star now appears as a classical DB white dwarf. Accretion may convert some of these stars into DBA white dwarfs.

As the star continues to cool, the He convection zone grows deeper, reaching its maximum depth near an effective temperature of about 13,000 K. At this point it dredges up traces of carbon from the composition-transition zone extending upward from the C/O core into the He convection zone, and the star becomes a DQ white dwarf. The base of the convection zone becomes degenerate when the effective temperature falls below about 13,000 K, and the increased efficiency of electron conduction begins to suppress convection in the deeper layers. This causes the base of the He convection zone to move back up toward the stellar surface. Diffusion then gradually drains the dredged-up C back out of the convection zone.

As the star cools further, the neutral He lines begin to fade, disappearing when the effective temperature drops below about 11,000 K. Accretion of interstellar matter, which continues in episodic fashion throughout the cooling of the white dwarf, continues to supply heavy elements to the star. The decline of other sources of opacity in the photospheric layers makes the traces of these accreted heavy elements observable in the cooler non-DA stars, and the white dwarf now is classified as spectral type DZ. Diffusion of course continues to cause the heavy elements to sink below the photosphere, and the observed abundances are determined by the balance between the rate of accretion and the rate of gravitational segregation. When the effective temperature drops below about 5,000 K, the star becomes too cool to display any absorption lines deeper than 10 % and is classified as a DC.

Astronomers thus have a plausible scenario to account for most of the spectral types of white dwarfs. Some difficulties still remain to be resolved, however, leaving yet another challenge for the future.

17. Music of the Spheres

Like many path-breaking events, the discovery of white dwarf pulsations happened serendipitously. In the mid-1960s, Arlo Landolt, an observational astronomer at Louisiana State University (LSU), was conducting photometric observations of some faint variable stars at the Kitt Peak National Observatory in the mountains near Tucson, Arizona. To help in counteracting variations in atmospheric transmission, Landolt performed his observations by first taking a photometric exposure on a target star and then switching to take a similar exposure on a nearby, faint star that was not variable. Alternating between the two stars every few minutes, Landolt built up a time series of observations that followed the variations in the target star's luminosity. In processing his data later, he planned to use the recorded observations of the non-variables to establish a steady baseline against which to measure the variability of his target stars.

This plan worked well except for one star, for which Landolt obtained results that made no sense. Being a careful scientist, he went back over every aspect of his observations and finally identified the problem. The white dwarf he had been using as a comparison standard for this particular star turned out to be a variable itself! This was a major surprise, because at the time "everybody knew" that white dwarf stars did not pulsate and, consequently, that they could safely be used as comparison standards.

Even more peculiar was the timescale of the variations, which roughly fit a sine curve with a period of approximately 12.5 min, or about 750 s.[1] This seemed completely mysterious, because the expected pulsation periods of white dwarfs were only some several seconds. These expectations were based on a theoretical computation of the (radial) oscillation periods of zero-temperature,

[1] Landolt, A. U. 1968, *Astrophys. J.*, **153**, 151, "A New Short-Period Blue Variable."

H.M. Van Horn, *Unlocking the Secrets of White Dwarf Stars*,
Astronomers' Universe, DOI 10.1007/978-3-319-09369-7_17

Chandrasekhar models done by Mme. Sauvenier-Goffin in 1949[2] and corrected a dozen years later by Evry Schatzman.[3] Similar calculations done in the 1970s showed that the period of the fundamental radial pulsation mode of, e.g., a 0.6 M_{Sun} white dwarf was about 13 s, consistent with a rough estimate of the time required for a sound wave to travel across the radius of a white dwarf.

Because white dwarfs are so faint and the expected values for the radial pulsation periods were so small, searches for oscillations in these stars had to wait on developments in technology. By the late 1960s, advances in electronics and photodetectors, coupled with the invention of the Fast Fourier Transform computer algorithm,[4] had made searches for rapid variability in stars practicable. Observers at Princeton University were quick to take advantage of this, but by the end of the decade their investigations had found no significant evidence for luminosity variations in white dwarfs over the period range from about 2–360 s. These null results, combined with the understanding of white dwarfs as cooling stars devoid of nuclear energy sources, led to the implicit assumption by astronomers during this early period that white dwarfs were stable, non-pulsating stars.

By the time of a conference on low-luminosity stars at the University of Virginia in 1968, however, Landolt had established that his white dwarf comparison star—an object called Haro-Luyten Taurus 76, meaning that it was the 76th white dwarf candidate identified by Haro and Luyten in the constellation Taurus—was indeed a variable.

As soon as Landolt reported this discovery,[5] astronomers immediately set to work to learn more about this interesting object. Did it have special properties that caused the pulsations? Were there other stars like it, or was it unique? The answer to the second question came a few years later, when in 1971 Barry Lasker and Jim Hesser, then at the Cerro Tololo Interamerican

[2] Sauvenier-Goffin, E. 1949, *Ann. d'Astrophys.*, **12**, 39, "La Stabilité Dynamique des Naines Blanches."

[3] Schatzman, E. 1961, *Ann. d'Astrophys.*, **24**, 237, "Sur la Période de Pulsation Radiale des Naines Blanches."

[4] Cooley, J. W., and Tukey, J. W. 1965, *Math. Comput.*, **19**, 297, "An Algorithm for the Machine Calculation of Complex Fourier Series."

[5] Landolt (1968), *op. cit.*

Observatory (CTIO), discovered pulsations in the white dwarf star R548 (WD 0133-116).[6] It turned out that this object had not only been identified as a white dwarf and given the name R548, but also that it had independently been identified as a variable star long before and had been given the variable star name ZZ Ceti. Because of the existing variable star designation, this star became the type member of the new class of pulsating white dwarfs.

As Lasker and Hesser found, the light curve of this star is not a simple sinusoid but displays clear beating between two modes with periods of approximately 213 and 274 s. However, it also displayed puzzling night-to-night amplitude variations, which frustrated early attempts to understand the oscillations. This problem was resolved in 1976, when E. L. ("Rob") Robinson, R. Edward Nather, and their graduate student at the time, John T. McGraw, carried out a series of high-speed photometric observations of this star, with overlapping data coverage extending from the Sutherland Observatory in South Africa to McDonald Observatory in Texas. With the 8-h difference in longitude between the two sites, the observers were able for the first time to resolve the oscillation spectrum completely. They found that what had been previously interpreted as two amplitude-variable oscillations with periods of 213 and 274 s were in fact two closely spaced *pairs* of highly *stable* modes.[7]

By 1980, sufficient data had been accumulated—including further significant contributions from CTIO—to enable astronomers to determine the periods of these modes with high precision. The results showed that the close spacing of each pair could be explained in terms of about a 1.5-day rotation period of the white dwarf.[8] Furthermore, each of the four separate oscillation periods proved to be stable to better than 1 s in more than 3,000 years.

What could these mysterious oscillations be? It did not take long for Ganesh Chanmugam (1939–2004), an astrophysicist at LSU[9], and for Brian Warner and Rob Robinson at the University of

---

[6] Lasker, B. M., and Hesser, J. E. 1971, *Astrophys. J.*, **163**, L89, "High-Frequency Stellar Oscillations. VI. R548, A Periodically Variable White Dwarf."

[7] Robinson, E. L., Nather, R. A., and McGraw, J. T. 1976, *Astrophys. J.*, **210**, 211, "The Photometric Properties of the Pulsating White Dwarf R 548."

[8] *Ibid.*

[9] Chanmugam, G. 1972, *Nature Phys. Sci.*, **236**, 83, "Variable White Dwarfs."

Texas,[10] to suggest that the long-period oscillations in the ZZ Ceti stars were *g*-mode oscillations of the white dwarf. In 1941, British astrophysicist Thomas G. Cowling had already shown that the pulsations of any star fall into two main groupings: the *p*-modes, for which the restoring force is provided by the pressure of stellar matter, and the *g*-modes, for which it is provided by gravity acting upon the pulsation-induced buoyancy of a mass element.[11]

The *p*-modes are essentially acoustic waves, with higher overtones having progressively shorter periods. However, the *g*-modes in white dwarfs all have much longer periods than the *p*-modes, and the higher overtones have progressively greater periods. In 1969, Annie Baglin and Evry Schatzman estimated the *g*-mode periods of white dwarfs with central temperatures of about 10^7 K, finding them to be of the order of hundreds of seconds.

The first detailed calculations of the non-radial oscillation modes in white dwarfs were carried out a few years later, in 1973, by Carl Hansen (1933–2011; see Figure 17.1), a faculty member at the Joint Institute for Laboratory Astrophysics (JILA) at the University of Colorado in Boulder, and Yoji Osaki, a JILA Visiting Fellow then on leave from the University of Tokyo.[12] They used a number of finite-temperature models for artificial pure-iron white dwarfs selected from stellar evolution sequences that Malcolm Savedoff and I had computed several years earlier at the University of Rochester. A senior thesis student at Rochester had performed some approximate calculations of the radial pulsation periods for a few models from these sequences—essentially confirming with digital computer calculations Schatzman's earlier results.

I had reported our results at a meeting of the American Astronomical Society at the University of Colorado, where I met Carl Hansen for the first time. He was interested in applying his powerful stellar-pulsation code to these models, so I sent him several models on decks of IBM cards when I returned to Rochester, and

[10] Warner, B., and Robinson, E. L. 1972, *Nature Phys. Sci.*, **239**, 2, "Non-Radial Pulsations in White Dwarf Stars."

[11] Cowling, T. G. 1941, *Mon. Not. Roy. Astron. Soc.*, **101**, 367, "Non-Radial Oscillations of Polytropic Stars."

[12] Osaki, Y., and Hansen, C. J. 1973, *Astrophys. J.*, **185**, 277, "Nonradial Oscillations of Cooling White Dwarfs."

FIGURE 17.1 Carl J. Hansen. Photograph taken in 1984 by Steven D. Kawaler. Reproduced with permission

these were the models Hansen and Osaki used in their work. Theirs was one of the first papers to show that the observed pulsation periods of the ZZ Ceti stars (pulsating white dwarfs with pure hydrogen atmospheres) were consistent with non-radial *g*-mode pulsations of these stars.

By the mid-1970s, it had become apparent that all the variable white dwarfs discovered by that time were stars with H-rich surface layers (denoted as DAVs, with "V" meaning "variable") and with effective temperatures in the narrow range from 10,000 K to about 13,000 K. The locations of these stars in the H-R diagram suggested the existence of a "white dwarf instability strip" that appeared to be an extension of the well-established Cepheid pulsational instability strip into the domain occupied by the degenerate stars. The Cepheid variables are massive Main Sequence stars that undergo radial pulsations driven by an instability in their subsurface partial-ionization zones.

At IAU Colloquium 53, held at the University of Rochester in 1979, Rob Robinson gave a review talk about the observational properties of the ZZ Ceti stars.[13] He emphasized (1) that all of the ZZ Ceti stars were perfectly ordinary DA (H-atmosphere) white dwarfs, except that they exhibited multi-modal pulsations; (2) that they were confined to a relatively narrow instability strip centered around photometric color $B\text{-}V = 0.20$; (3) that the pulsation periods were very long (ranging from hundreds to thousands of seconds, in contrast to the calculated radial pulsation periods of several seconds); and (4) that the periods were consistent with *g*-mode pulsations. However, it was not then known what drove these pulsation modes or why only certain modes out of the larger number of possibilities were excited.

After Rob's talk, Carl Hansen and I sat on a stone bench in the middle of the quadrangle on the Rochester campus and discussed these observational constraints with Don Winget (Figure 17.2), who was then working on his Ph.D. thesis at Rochester, computing theoretical models for white dwarf pulsations under the joint supervision of Carl and myself. I no longer recall who first came up with the idea, but we all agreed that perhaps we should pay more attention to the spectroscopists' emphasis on the purity of the hydrogen surface layers in these stars. It had long been known that the surface gravities of white dwarfs are so high that heavier elements rapidly settle out of the atmosphere. Maybe we should start thinking about investigating the pulsation properties of white dwarf models having stratified surface layers, with a pure H layer atop a pure He layer, which in turn sits atop a pure C core.

That decision ultimately proved to be the key to our understanding of white dwarf pulsation modes. Don was eager to tackle the project, and we decided to use a grid of stratified models, with varying H/He/C compositions, a number of different masses, and a range of surface temperatures that more than spanned the observational instability strip. At the same time, Wojtek Dziembowski in Poland and Gerard Vauclair in France, both established experts in the field of stellar oscillations, were hot on the trail of the pulsating DA white dwarfs, and Don felt the competition keenly.

Mathematically, a stellar oscillation mode is represented as a function that depends upon the radial distance from the center of

[13] Robinson, E. L., in Van Horn and Weidemann (1979), p. 343, "The Observational Properties of the ZZ Ceti Stars."

FIGURE 17.2 Don Winget (background) at the University of Texas in 1985. Photograph taken by Steven D. Kawaler during a visit by Icko Iben, Jr. (foreground). Reproduced with permission

the star multiplied by a function of latitude and longitude. The radial function has some number of nodes (radial locations where the function goes through zero), represented by an integer k. A mode with $k=0$ has no nodes in the radial function; one with $k=1$ has a single node somewhere between the center and the surface; and so on. The higher the value of k, the more "wiggly" is the function.

Similarly, the functions of latitude and longitude—called "spherical harmonics"—have nodes represented by the integers l and m. A spherical harmonic with $l=0$ and $m=0$ has no nodes. One with $l=1$ has a single node (the "equator") separating the stellar surface into two hemispheres ("north" and "south); this is called a "dipole" mode. A mode with $l=2$ is a "quadrupole" mode. Non-zero values of m correspond to nodes in longitude. Figure 17.3 illustrates the patterns described on the stellar surface

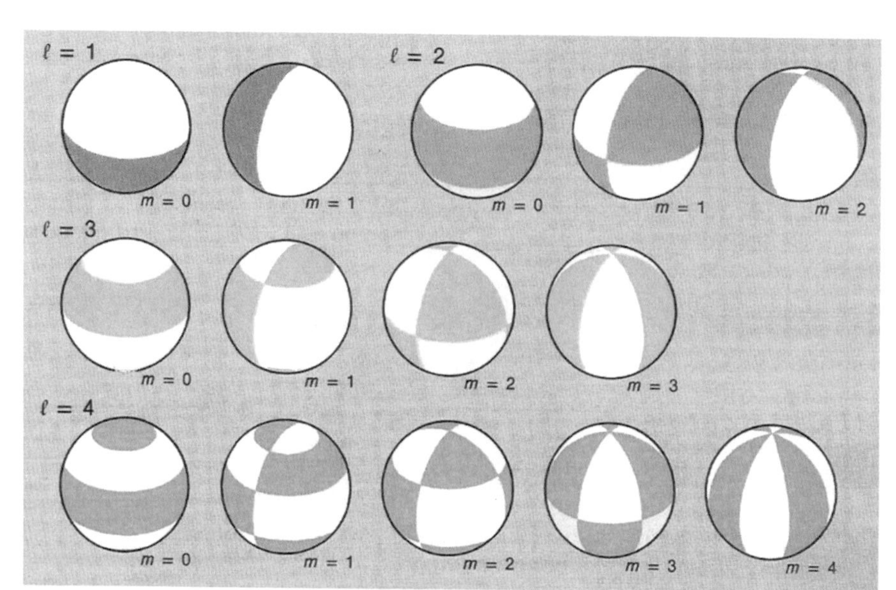

FIGURE 17.3 Angular patterns of some non-radial pulsation modes. The colored patches can be thought of as regions of cooler surface temperature and the white patches as regions of higher surface temperature. During the course of an oscillation, the high-temperature regions become low-temperature ones and *vice versa*. Values of the indices l and m that characterize each mode are indicated. From Winget and Van Horn 1982. Reproduced with permission

by spherical harmonics with a number of different values of l and m. All of these properties of the stellar oscillation modes come into play in seeking to associate periodicities observed in a star's light curve with the underlying physical oscillation modes.

In the course of his Ph.D. thesis calculations,[14] Don discovered that some oscillations in the stratified white dwarf models had appreciable amplitudes only in the thin surface hydrogen layers.[15] We termed these "trapped modes." Other modes had significant amplitudes distributed throughout the body of the star.

[14] Winget, D. E., and Van Horn, H. M. 1982, *Sky & Tel.*, **64**, 216, "ZZ Ceti Stars: Variable White Dwarfs."

[15] Similar conclusions about the effects on white dwarf pulsations of stratified surface layers were reached by Dziembowski, W., and Koester, D. 1981, *Astron. & Astrophys.*, 97, 16, "Excitation of Gravity Modes in White Dwarfs with Chemically Stratified Envelopes," and by Dolez, N., and Vauclair, G. 1981, *Astron. & Astrophys.*, **102**, 375, "Gravity Modes in ZZ Ceti Stars. I. Quasi-Adiabatic Analysis of Overstability."

Don discovered further that the thickness of the hydrogen layer and the temperature of the star determined which specific modes were trapped. For the first time, this offered the possibility of determining constraints on the thicknesses of the superficial H and He layers by matching the oscillation periods observed in specific white dwarfs against a list of trapped modes from models with similar masses and effective temperatures. For stratified models in which the outermost layer consists of hydrogen, Winget's calculations produced results that were generally consistent with the observed properties of the ZZ Ceti stars, though the models were not yet sufficient to provide specific mode identifications.

In addition, Don found the reason why only a few oscillation modes are excited in the pulsating white dwarfs. The inner parts of the star act to damp the oscillation modes, while the partial-ionization zones near the stellar surface provide the excitation mechanisms responsible for driving the oscillations to observable amplitudes, just as is the case for other pulsating stars that had been studied for decades. The difference is that the surface partial-ionization zones in white dwarfs are very much thinner than is the case for other variable stars. Consequently, for a mode with significant amplitude distributed throughout a white dwarf, the damping provided by the deep interior outweighs the driving by the thin surface layer, so that the mode is not excited. Conversely, for the trapped modes, which have appreciable amplitudes only in the thin surface layers, the excitation mechanisms operating in the partial-ionization zones can succeed in overcoming the damping so that the oscillation can grow large enough to be observable.

By 2007, some 126 DAV stars had been discovered,[16] with a mean mass of 0.58 $M_{\rm Sun}$, in excellent agreement with the mean mass of all DA white dwarfs. The oscillation periods range from a few hundred seconds up to more than a thousand seconds, and all are multi-mode pulsators. The fact that the colors (and thus effective temperatures) of these stars are so similar provides a clear indication that the cause of the pulsations is something inherent in the stars themselves. The hottest DAV has an effective temperature of approximately 12,460 K, which effectively defines the

[16]Kepler, S. O. 2007, *Comm. In Asteroseismology*, **150**, 221, "Observational White Dwarf Sesimology."

"blue edge" of the ZZ Ceti instability strip, while the coolest has an effective temperature of about 11,160 K, which defines the "red edge." The hotter stars tend to have shorter periods and fewer excited modes, while the cooler DAVs have longer periods and more complicated oscillation spec tra.

Winget computed some additional models with surface layers consisting of helium rather than hydrogen. For these DB models, he found the very interesting result that similar trapped modes were excited by the helium partial-ionization zones, though at significantly higher effective temperatures than for the ZZ Ceti models. Accordingly, in 1982 he predicted the existence of a pulsational instability strip for the DB (He-atmosphere) white dwarfs, which he estimated to occur near an effective temperature of about 20,000 K.[17] This prediction flew in the face of the prevailing opinion that DB white dwarfs did not sustain pulsations.

Upon finishing his dissertation, Winget accepted a postdoctoral appointment with professors Nather and Robinson at the University of Texas. Both men were experts in the high-speed photometry of variable white dwarfs and other stars, and Don hoped to persuade them to help him search for pulsating DB white dwarfs with observed colors similar to those corresponding to the effective temperatures for which his theoretical calculations had predicted observable oscillations. They concurred and helped him set up an observing program at the University's McDonald Observatory in west Texas to look for the predicted variables.

It was a very satisfying confirmation of the theory when on the night of May 26, 1982, non-radial oscillations were discovered in the DB white dwarf GD 358 (WD 1645+325).[18] To my knowledge, this was the first time the existence of a new class of variable star had been predicted theoretically prior to its discovery.

The morning after the discovery of the first pulsating DB, I was working in my office at the University of Rochester when the telephone rang. When I answered, I heard the voice of my former student, Don Winget, bleary from lack of sleep but sky-high with excitement. "Are you sitting down?" he asked. When assured

---

[17] Winget, D. E., *et al.* 1982, *Astrophys. J.*, **252**, L65, "Hydrogen-Driving and the Blue Edge of Compositionally Stratified ZZ Ceti Star Models."

[18] Winget, D. E., *et al.* 1982, *Astrophys. J.*, **262**, L11, "Photometric Observations of GD 358: DB White Dwarfs Do Pulsate."

that I was, he continued, "We found a pulsating DB last night!" That ratcheted up my excitement, too, and after congratulating him on the discovery, I demanded to know all the particulars. It turned out that the star was hotter than his initial estimates had been, but its pulsation properties were otherwise very similar to those of the pulsating DA white dwarfs—it displayed multiple oscillation modes with periods of hundreds of seconds. Before ringing off to make the several-hundred mile return journey from McDonald Observatory to the university in Austin, Don remarked that "Much beer was consumed on the mountain last night!" in celebration of the discovery.

The discovery of the pulsating DB white dwarfs had three important ramifications: (1) It confirmed the basic understanding of the properties of non-radial oscillations in white dwarfs. (2) It confirmed that the surface layers of white dwarfs are compositionally stratified. (3) It raised the possibility of utilizing the non-radial oscillations of white dwarfs to conduct "asteroseismological" investigations of their interiors.

Most of the DB white dwarfs examined in the survey that led to the discovery of pulsations in GD 358 proved not to be pulsators. However, as of 2007, 13 DB variables (DBVs) were known.[19] As expected, their oscillation spectra resemble those of the ZZ Ceti variables, with periods ranging from about 150 s to 1,000 s.

The prototype DB variable GD 358 (WD 1645+325, also known as the variable star V477 Herculis), has been the subject of extensive investigations. In their 1982 discovery paper,[20] Winget and his colleagues found that at least 28 oscillation modes in the period range from 142.3 to 952 s are excited simultaneously in this star, with amplitudes ranging up to 0.30 magnitudes. Many of the pulsations fall into groups of four or five modes equally spaced in frequency, suggesting that they are rotationally split $l=2$ modes and implying a rotation period of about an hour and a half.

Soon after the discovery of pulsations in GD 358, the temperature of this star was found to be considerably higher than Winget's original theoretical estimate of 20,000 K. Several groups of astronomers made independent observational temperature

[19] Kepler (2007), *op. cit.*

[20] Winget *et al.* (1982), *op. cit.*

determinations over the next decade and a half, using UV spectra obtained from spacecraft as well as ground-based optical spectra. In 1999 Alain Beauchamp and his colleagues found the effective temperature of this star to be 24,900 K by analyzing the observational data using Beauchamp's then-new pure helium model atmospheres.[21] The discrepancy between Winget's original theoretical model and the observational values for the stellar temperature was resolved by using a more efficient version of the mixing-length theory of convection in constructing a model for this star.

Observers also worked to determine the red and blue edges of the DBV instability strip. Beauchamp and his colleagues also used his pure-He model atmospheres to analyze the observations, finding the blue edge to lie between 22,400 and 27,800 K.[22] In contrast to the DA variables, however, convective mixing is not an attractive hypothesis to explain the red edge, since mixing with the underlying carbon core would lead to heavily carbon-dominated atmospheres, which have not been observed in these stars.

A third class of variable degenerate stars consists of pulsating stars that are *much* hotter than the DBVs. The DOV star PG 1159-035 was the first member of this class to be discovered. (The DOV stars were actually the *second* class of variable degenerates to be discovered chronologically; the DBV stars were third, with GD 358 being the first representative to be discovered.) PG 1159-035 is one of a large number of very blue (and thus very hot) stellar objects found in an extensive survey conducted at the Palomar Observatory by Richard Green in 1977.[23]

Because this star displays a very blue continuum accompanied by He II emission lines, it was originally thought to be a possible He-mass transfer binary like AM Canis Venaticorum. It was therefore expected to exhibit "flickering" in intensity associated with the mass transfer. For this reason, John McGraw and his collaborators undertook high-speed photometry of this star. On April 29, 1979, observations with the University of Arizona's Steward Observatory Multiple Mirror Telescope (MMT) revealed

[21] Beauchamp. A., *et al.* 1999, *Astrophys. J.*, **516**, 887, "Spectroscopic Studies of DB White Dwarfs: The Instability Strip of the Pulsating DB (V777 Her) Stars."

[22] *Ibid.*

[23] Green (1977).

multiperiodic oscillations in this star, with periods of about 539 and 460 s.[24]

Just as had happened with the ZZ Ceti stars, the type member of this new class of pulsating stars had been identified earlier as the variable star GW Virginis. This class includes both stars like PG 1159-035 and the closely related variable central stars (or "nuclei") of planetary nebulae, called "planetary nebula nucleus variables," or PNNVs. The temperatures of all these stars are very high. In addition to their very blue colors, direct evidence for the high temperatures comes from two other sources. The spectra of these stars are dominated by absorption lines of highly ionized species such as He II, C IV, and O VI.[25] In addition, Martin Barstow and his colleagues in 1986 detected photospheric X-ray pulsations from PG 1159-035 from *EXOSAT*,[26] which indicated an effective temperature in excess of 100,000 K.

A fourth class of short-period variable stars was discovered in 1997—pulsating hot subdwarfs of spectral class B, called "sdBVs." Although the hot subdwarfs also are thought to be precursors of some white dwarfs, like the very hot central stars of planetary nebulae, their pulsation characteristics are quite different from those of the oscillating white dwarfs, and we shall not discuss them further.

What excites the observed pulsations in all these variable degenerate stars? Although the specific physical processes are different for each of the three classes, they have in common the same general mechanisms. Of course, whether or not a particular star experiences a net gain or loss of energy during the course of a pulsation cycle depends upon the balance between the driving provided by the excitation regions and the damping provided by other regions of the star.

One suggested excitation mechanism—originally identified decades earlier as one of the causes of oscillations in the Cepheid

---

[24]McGraw, J. T., *et al.*, in Van Horn and Weidemann (1979), p. 377, "PG 1159-035: A New, Hot, Non-DA Pulsating Degenerate."

[25]Wesemael, F., Green, R. F., and Liebert, J. W. 1985, *Astrophys. J. Suppl.*, **58**, 379, "Spectrophotometric and Model-Atmosphere Analyses of the Hot DO and DAO White Dwarfs from the Palomar-Green Survey."

[26]Barstow, M., *et al.* 1986, *Astrophys. J.*, **306**, L25, "The Detection of Photospheric X-Ray Pulsations from PG1159-035 with *EXOSAT*."

variable stars—is an opacity effect. Often called the "kappa mechanism," after the usual astrophysical symbol for opacity, this process operates in a region of incomplete ionization. During the compression phase of a pulsation, the opacity increases, impeding the flow of energy out of a mass element. Consequently, that element has relatively more energy than adjacent regions, which is then available to augment the subsequent expansion phase.

A second proposed mechanism, which also operates in the Cepheid variables, is called the "gamma mechanism," after the symbol used to relate the variation of pressure with density upon compression. This effect also operates in a partial-ionization zone, where compression causes energy to be stored as internal energy of the partially ionized species rather than being dissipated as heat. Consequently, this energy also is available to augment the amplitude of the oscillation during the subsequent expansion phase of the pulsation cycle.

Another excitation mechanism, originally identified by Thomas Cowling in a different context, was first investigated in connection with white dwarf oscillations in 1983 by A. J. Brickhill.[27] He pointed out that, although earlier calculations for the pulsating white dwarfs—including those done by Winget, Hansen, and this author—had assumed that the convective motions remain "frozen" during the course of a pulsation cycle, in fact the opposite is true. The timescale for convective turnover is much shorter than the pulsation period, making this assumption untenable.

To explore this regime, Brickhill carried out a time-dependent calculation of the amplitude of the surface luminosity variation and the corresponding dissipation rate using the mixing-length theory of convection. He found that many modes are excited simultaneously, up to some maximum period, and that the maximum period increases as the white dwarf cools, in agreement with observations. This mechanism of mode excitation, which he termed "convective driving," results from the fact that the convection zone functions like a heat engine, absorbing energy when compressed and releasing it when expanded.

---

[27] Brickhill, A. J. 1983, *Mon. Not. Roy. Astron. Soc.*, **204**, 537, "The Pulsations of ZZ Ceti Stars."

In addition, Brickhill confirmed the 1981 discovery by Noel Dolez and Gerard Vauclair that a mode is stable when the oscillation period is longer than a particular characteristic timescale over a large fraction of the driving zone.[28] Accordingly, Brickhill concluded that this stabilization of the longer-period modes at lower effective temperatures is the mechanism responsible for producing the red edge of the ZZ Ceti instability strip, rather than convective mixing, and the attendant dilution of the surface hydrogen.

In 1998 Yanqin Wu considered Brickhill's "convective driving" theory in substantially greater detail.[29] She and her Caltech thesis adviser, Peter Goldreich, agreed with Brickhill that the convection zone is the location of mode excitation in the ZZ Ceti stars, because this region bottles up the perturbed heat flux entering it from below, and their calculations support most of Brickhill's conclusions. Largely on the basis of their work, which also called renewed attention to Brickhill's earlier calculations, astronomers now generally accept convective driving as the excitation mechanism for white dwarf oscillations, at least for the DAV and DBV stars. One of Wu's and Goldreich's principal results was the discovery that the amplitudes of the *g*-modes in the ZZ Ceti stars are determined by the nonlinear transfer of energy from the observed modes through a parametric instability. This occurs when the amplitude of a mode exceeds a certain critical value, when two "daughter" modes are excited. The daughter modes, however, are strongly damped, draining energy from the parent mode "cataclysmically" and halting the growth of the parent mode at this amplitude.

Thus, by the end of the twentieth century, astronomers had gained a pretty clear understanding of the variable white dwarfs. The next step was to employ the pulsations of these stars to determine some of their physical properties.

---

[28] Dolez, N., and Vauclair, G. 1981, *Astron. & Astrophys.*, **102**, 375, "Gravity Modes in ZZ Ceti Stars. I. Quasi-Adiabatic Analysis of Overstability."

[29] Wu (1998); see also Goldreich, P., and Wu, Y. 1999a, *Astrophys. J.*, **511**, 904, "Gravity Modes in ZZ Ceti Stars. I. Quasi-Adiabatic Analysis of Overstability;" Wu, Y., and Goldreich, P. 1999, *Astrophys. J.*, **519**, 783, "Gravity Modes in ZZ Ceti Stars. II. Eigenvalues and Eigenfunctions;" and Goldreich, P., and Wu, Y. 1999b, *Astrophys. J.*, **523**, 805, "Gravity Modes in ZZ Ceti Stars. III. Effects of Turbulent Dissipation."

18. The Whole Earth Telescope and Asteroseismology

With the discovery of variability in several different types of degenerate stars, and with a growing understanding of their pulsation properties, astronomers began to employ the multitude of oscillation modes to learn more about these objects, a practice that has come to be called "asteroseismology."

Initially, progress in analyzing the pulsations of white dwarfs was hampered by a seemingly insurmountable obstacle, as ground-based observations of these faint stars can be carried out only at night. But no matter how many consecutive nights were devoted to continuous photometric observations of a given star, there seemed to be no way to connect this time series across the daylight hours to form a single, consecutive record. In consequence, the spectrum of oscillations in the time-varying light curve of the star could not be resolved sufficiently well to enable an astronomer to tell what frequencies were actually present.

To determine the spectrum of oscillation frequencies in a pulsating white dwarf, astronomers pass the light curve through a mathematical operation called the Fourier transform, using a digital algorithm called the Fast Fourier Transform, or FFT.[1] This produces a distribution of peaks at different frequencies, called a power spectrum, with the tallest peaks representing the most prominent periodicities in the spectrum. However, the widths of the peaks depend on the length of the light curve being analyzed, with the spectrum corresponding to a single night's observations having moderately broad peaks that can hide smaller secondary features. Even when many nights are analyzed together, the gaps

[1] Cooley, J. W., and Tukey, J. W. 1965, *Math. Comput.*, **19**, 297, "An Algorithm for the Machine Calculation of Complex Fourier Series."

H.M. Van Horn, *Unlocking the Secrets of White Dwarf Stars*, Astronomers' Universe, DOI 10.1007/978-3-319-09369-7_18

in the data produced by the daylight hours act as an additional periodicity, which can interact with the real signal to produce a "forest" of spurious "aliases" at other frequency points.

With a multi-periodic signal of the sort produced by the variable white dwarfs, it is thus difficult or even impossible to distinguish from such data which frequencies correspond to photometric oscillations of the star and which to diurnal aliases of other frequencies. For example, the slow rotation of a white dwarf—with a rotation period P_{rot} of the order of hours to days—produces rotationally split components of the stellar oscillation modes, which have frequency spacings similar to those of the diurnal aliases. In addition, every observational data record is subject to "shot noise" resulting from the limited numbers of photons detected and to unwanted variations produced by atmospheric scintillations. These sources of error in the data affect the power spectrum, so that the highest peak in a noisy signal may not always correspond to the dominant periodicity. The only way to achieve adequate resolution and reduce or eliminate the aliases is to get rid of the interruptions caused by the daylight hours. But how could this be accomplished?

The answer to this question was provided by an inspiration from Ed Nather at the University of Texas (1926–2014; Figure 18.1).[2] Nather did not start out to pursue a career in astronomy. Instead, at the end of World War II, he spent 2 years as an electronics technician for the U. S. Navy. After the war, he returned to college, graduating with most of his courses in English. He then took a job at the Hanford nuclear facility in the state of Washington, capitalizing on the technological expertise acquired during his Navy service.

One of Nather's tasks at Hanford was the scary job of helping to service the core of a nuclear reactor. This required him to don protective clothing, and even then he could only remain inside the reactor core for a few minutes at a time to avoid excessive exposure to dangerous levels of radiation. During the 1960s, Ed became fascinated with the newly developing computer technology, and he designed and wrote FORTRAN compilers for some small com-

[2] Solheim, J.-E. 1993, *Baltic Astron.*, **2**, 363, "R. E. Nather—The Founder of the WET."

FIGURE 18.1 R. Edward Nather. Recent image courtesy David Nather. Reproduced with permission

puting firms before ending up as the technical director of the Nuclear Instrumentation Division at Beckman Instruments.

At this point the University of Texas succeeded in attracting him to join the astronomy department as a senior technician. There his knowledge of the design and programming of digital computers enabled him to become a leader in the development of high-speed astronomical photometry, which was then starting to take off. It soon became clear, however, that Ed's intellectual capabilities were not being adequately utilized as a technician and that the path to a more productive and satisfying career required a doctorate. Accordingly, at a more mature age than is the norm, Nather undertook studies leading to a Ph.D. in astronomy working under the supervision of Brian Warner, who held appointments both at the University of Texas and at the University of Cape Town in South Africa and who is himself a leading expert in high-speed astronomical photometry. Nather received his degree in 1972 and was immediately appointed to the astronomy faculty at the University of Texas, where he remained for the rest of his career. Together with Warner and E. L ("Rob") Robinson, he helped build one of the world's leading groups in the field of high-speed photometry, guiding generations of students in the development and astronomical use of sophisticated instrumentation, employing his superb understanding of the English language in helping them learn to write clearer and more interesting scientific papers, and in analyzing vast amounts of astronomical data to extract the maximum amount of information possible and thus greatly increasing our knowledge of a variety of astronomical objects.

Nather's insightful idea for eliminating the daylight interruptions in the photometric record of a pulsating white dwarf was to organize closely coordinated observations at a sequence of observatories around the globe, enabling each observatory at local dawn to "hand off" the responsibility for continuing high-speed photometry of a particular target star on to another observatory farther west in longitude, where it was still nighttime. Indeed, astronomers from Texas and South Africa had previously done exactly that in resolving the light curve of ZZ Ceti itself.[3]

If enough observatories agreed to participate, a target star could be handed along again and again, building up the equivalent of a single, continuous time series of observations that could be days or weeks in length. And such a long data stream would indeed have the resolution needed to identify the precise frequencies present in the light curve of the target star. Nather termed this concept the "Whole Earth Telescope," or WET,[4] and he immediately set about organizing the logistics to accomplish this, including the equipping of participating observatories with identical high-speed photometers in order to facilitate the data analysis. Figure 18.2 shows the distribution of observatories that participated in a WET campaign in March 1989.

Observing campaigns for the Whole Earth Telescope—called "extended coverage" or "XCOV" campaigns—are planned in advance and are coordinated in real time from a single headquarters site. In this way, the consortium of participating observatories acts as much as possible like a single instrument. Targets are assigned for each site based on previously established scientific priorities, local weather patterns, and overlapping longitude coverage, with different observatories participating in different campaigns. Priority is given to targets visible from both northern and southern hemispheres in order to increase coverage and minimize lost time.

[3] Robinson, E. L., Nather, R. A., and McGraw, J. T. 1976, *Astrophys. J.*, **210**, 211, "The Photometric Properties of the Pulsating White Dwarf R 548."

[4] Nather, R. E., in Wegner (1989), p. 109, "The Whole Earth Telescope;" Nather, R. E., *et al.* 1990, *Astrophys. J.*, **361**, 309, "The Whole Earth Telescope: A New Astronomical Instrument."

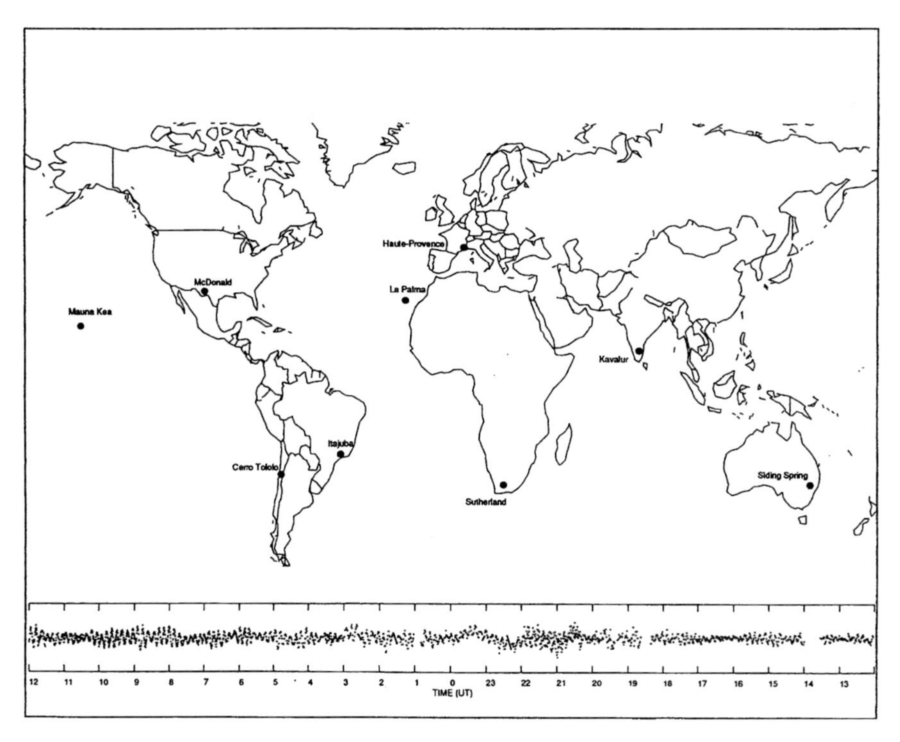

FIGURE 18.2 The distribution of observatories that participated in the WET campaign in March 1989, for which the principal target star was PG 1159-035. The "wiggly" curve along the bottom of the figure is a section of the observed light curve for this star, recorded as a function of Universal Time (UT). From Nather *et al.* 1990. © AAS. Reproduced with permission

At each location, observers obtain simultaneous photometric time-series data for their designated target star, a comparison star, and sky background using a high-speed "three-channel" photometer. At local dawn, the data are sent by e-mail to the WET headquarters site, which maintains long-distance telephone contact with each of the participating observatories, and the data are reduced and analyzed there. The target star is then handed off to another WET observing site farther west, and the process is repeated around the planet. To paraphrase a well-known quotation, "The Sun never rises on the Whole Earth Telescope."

The WET campaigns were initially coordinated from Nather's home institution, the University of Texas (Austin), later from Iowa State University by Steven D. Kawaler, one of Don Winget's former students, and are currently coordinated from the University of Delaware by Judi Provencal, another Winget protégé. A particular advantage of this central coordination is that real-time data analysis permits the adjustment of observing priorities right at the telescope, a feature that has proven quite valuable. Since the first observing run in March 1988, up to three XCOV campaigns have been conducted every year.

In order to combine the time-series data from the WET consortium into a single homogeneous dataset, it is helpful to have as much uniformity as possible in hardware, software, and observing procedures at the different sites. The initial data-acquisition system intended for use by the participating observatories was a portable two-star photometer based on Nather's original 1973 design.[5] A three-channel system designed in the mid-1990s also allows continuous measurement of the sky background and permits observations to be extended significantly into the regime of astronomical twilight (20 min more each morning and evening). The photometer is controlled by a laptop computer that records the data and displays the accumulating light curve, allowing real-time monitoring by the observers. Because funds sufficient to equip all the participating observatories with identical instruments never became available, WET observations have in reality been carried out using a multitude of different two- and three-channel photometers of differing designs. This leads to difficulties with the WET that are not easily removed. For example, (1) the use of telescopes with different apertures produces differences in the signal-to-noise ratio in different portions of the combined record, and (2) the atmospheric extinction at different sites complicates the concatenation of data.

The process of utilizing the observed spectrum of oscillation modes in conjunction with detailed theoretical model calculations to determine the physical properties of a star has been named "asteroseismology." This has proven to be particularly useful for the pulsating white dwarfs. They tend to exhibit large numbers of oscillation modes, which provide considerable leverage to astronomers working to extract values for different physical quantities from the observational data. If enough modes are excited and can

[5] Nather, R. E., in Beer (1973), p. 91, "High-Speed Photometry."

be identified, comparison between theory and observation can yield a wealth of astrophysical information about the star. Because asteroseismology depends only upon calculations of the pulsation-mode structures and frequencies—and *not* on the more sensitive details of mode excitation—it is free from any uncertainties associated with the actual driving mechanism of the oscillations.

As an illustration of the power of this approach, let us consider the asteroseismological analysis of the pulsating DOV star PG 1159-035, familiarly abbreviated as PG 1159. Prior to 1990, high-speed photometric observations had revealed that this star has scores of oscillation modes grouped into several different sets closely spaced in frequency. Accordingly, the Whole Earth Telescope consortium selected this star as the primary target for a WET campaign planned for the first two weeks of March 1989. For this sequence of measurements, observatories spanning longitudes all around the globe were involved, and the consortium provided 264 h of nearly continuous high-speed photometry for PG 1159. In their 1991 paper describing these observations,[6] Winget and his colleagues reported finding 125 individual modes with periods ranging from 385 s to 1,000 s.

In his 1986 University of Texas Ph.D. thesis,[7] Steve Kawaler described the asteroseismological analysis of such data. He began by noting that the observed oscillation modes of PG 1159 are high-overtone *g*-modes (large integer values of the radial index k), with low integer values of the spherical harmonic index l. For these high-order modes, consecutive radial overtones are spaced approximately equally in period, with the spacing depending primarily on the mass of the white dwarf. Kawaler showed how the observed period spacing can easily be compared with that calculated for theoretical models having different masses to determine the stellar mass. Following Kawaler's procedure, Winget and his colleagues in 1991 were able to obtain a value for the mass of PG 1159 of $(0.586 \pm 0.003)\ M_{Sun}$.[8]

[6] Winget, D. E., *et al.* 1991, *Astrophys. J.*, **378**, 326, "Asteroseismology of the DOV Star PG 1159–035 with the Whole Earth Telescope."

[7] Kawaler (1986).

[8] Winget, D. E., *et al.* (1991), *op. cit.*

The observed period spacings in PG 1159-035 are not strictly uniform, however. Kawaler's pulsation calculations showed that while *g*-modes probe the deep interior when a star is still at the extremely high temperatures characteristic of the central stars of planetary nebulae, they become concentrated in the non-degenerate envelope as the star cools down into the white dwarf domain. For models with mass 0.6 M_{Sun}, Kawaler found the transition between these two types of behavior to occur at about 100 times the luminosity of the Sun, which is roughly the luminosity of PG 1159-035. Further, while the envelopes of these high-gravity stars are becoming compositionally stratified, the process has not proceeded as far as it has in the much cooler white dwarfs. In any case, the composition does not change discontinuously, as Winget's earlier calculations had assumed, but rather varies from one composition to another through a "composition transition zone." The modes trapped in the surface layers of the PG 1159 stars probe these regions of varying composition, and this produces the observed differences from a uniform period spacing.

Following Kawaler's approach, the WET group used the observed departures from uniformity to constrain the compositional stratification in PG 1159-035.[9] They found that the outer envelope of the star definitely contains two layers differing in composition, but they were unable to tell what the different layers were. A few years later, Steve Kawaler and Paul Bradley showed that the observed period distributions actually match very well with standard models for post-AGB stars.[10] Their best fit corresponds to a model with mass $(0.59 \pm 0.01)\ M_{Sun}$, consistent with that determined earlier by Winget *et al.*, an effective temperature of about 136,000 K, and a surface He layer with a mass of 0.004 M_{Sun}.

As noted above, the oscillation modes of PG 1159-035 are grouped into sets of modes closely spaced in frequency. Most were found to be grouped into triplets with uniform frequency spacings, but some occurred in groups of four or five modes with somewhat larger uniform spacings. The most obvious way for such groupings

---

[9] *Ibid.*

[10] Kawaler, S. D., and Bradley, P. A. 1994, *Astrophys. J.*, **427**, 415, "Precision Asteroseismology of Pulsating PG 1159 Stars."

to be produced is by the mode-splitting caused by rotation of the star. As pointed out by Unno *et al.* in 1979,[11] the frequency of an oscillation mode of a rotating star, as measured in the rest frame of an observer, is split by the rotation rate[12] $\Omega = 1/P_{rot}$, where P_rot is the rotation period.

For a mode with given values of the integers *k*, *l*, and *m*, the shift in the oscillation frequency is equal to $m\Omega$ multiplied by a constant that depends on both the structure of the star and the indices *k* and *l*. A mode with a given value of *l* is split by the star's rotation into $(2l+1)$ modes with values of *m* ranging from $-l$ to $+l$. This immediately suggested to Winget and his colleagues that the triplets were $l=1$ modes, while the quintuplets were $l=2$ modes. In this way, they were able to assign spherical harmonic indices *l* and *m* to 101 of the observed modes and to show from the observed mode splittings that PG 1159-035 is rotating with a period of (1.38 ± 0.01) days.[13]

This rotation period is surprisingly slow. As a star evolves from the Main Sequence *en route* to becoming a white dwarf, it loses mass and presumably angular momentum as well. Unless angular momentum—the product of mass times radius times angular velocity (rotation rate)—is somehow transferred out of the stellar core that ultimately becomes the white dwarf, conservation of angular momentum would lead to much more rapid rotation in the white dwarf stage than is observed. This is the same physical principle that causes an ice skater's rate of spin to increase dramatically as she draws her arms in close to her body.

We can apply the same principle to a star, assuming arbitrarily that it rotates as a solid body on the Main Sequence, so that we can determine the rotation rate of the core from the observed rotation rate of the stellar surface. For a star with Main Sequence mass 2.0 M_{Sun}, which will become a white dwarf with a mass of about 0.6 M_{Sun}, the observed rotation period is a few days. If the central part of the star that is to become the white dwarf retains its original angular momentum throughout the remainder of its evolution, however, the expected white dwarf rotation would be very much

[11] Unno *et al.* (1979).

[12] The Greek letter Ω is frequently used as a symbol for the rotation frequency.

[13] Winget, D. E., *et al.* (1991), *op. cit.*

faster, with rotation periods ranging from seconds to minutes, rather than days. That white dwarf rotation periods are decidedly longer than this clearly demonstrates that most of the angular momentum must be transferred from the stellar core to the outer layers, which are lost. Exactly how and when this happens, however, still remains a challenge for astrophysicists to understand.

Since the original work on white dwarf pulsations in the late 1960s and early 1970s, it has been known that the periods of the *g*-modes depend upon the effective temperature of the star as well as its mass. Comparing the observed periods with those of models having different effective temperatures thus can in principle determine the stellar temperature. However, astronomers have found that observations and analyses of spectral lines tend to yield more accurate values for the effective temperature, and they have consequently tended to use these spectroscopic temperatures in their asteroseismological analyses.

With the mass of a degenerate star determined from asteroseismology and the effective temperature obtained from spectroscopy, an astronomer can obtain the intrinsic stellar luminosity from stellar evolution calculations. The luminosity and temperature yield the absolute visual magnitude M_V, and comparison with the measured apparent visual magnitude m_V then yields the "seismic" distance to the star. An excellent example of this approach was provided in 2000, when M. D. Reed, Steve Kawaler, and M. Sean O'Brien analyzed the composite-spectrum binary PG 2131+066.[14] This system consists of a PG 1159-type star and a low-mass Main Sequence star of spectral type M2.

From their seismic analysis of the PG 1159-type star, they obtained its mass as $M = (0.608 \pm 0.011)\ M_{Sun}$. With an effective temperature of 95,000 K determined from spectroscopy, Reed and his coworkers next found the luminosity of the degenerate star and the corresponding absolute visual magnitude. Comparison with the measured apparent magnitude of the PG 1159 star then gave a "seismic" distance of about 670 pc. This agrees reasonably well with the distance of about 560 pc obtained from the M2 com-

[14] Reed, M. D., Kawaler, S. D., and O'Brien, M. S. 2000, *Astrophys. J.*, **545**, 429, "PG 2131+066: A Test of Pre-White Dwarf Asteroseismology."

panion, providing both an important check on the asteroseismological method and a valuable calibration of this technique.

In addition, because the periods of the *g*-modes are sensitive to temperature, they gradually increase as a white dwarf cools and contracts. Observations of the rate of change of the pulsation periods in principle can thus constrain the white dwarf cooling rate, core composition, and neutrino emission rate. This has proven to be an exceptionally challenging task, however.

In 1985, Winget and several colleagues first measured the rate of change of a very stable 516-s periodicity in PG 1159-035 and found it to be *de*creasing by about 4 s per 10,000 years, corresponding to a timescale for changes in this mode of about 1.2 million years.[15] Steve Kawaler's theoretical calculations,[16] published a year later, agreed with the magnitude of this change, but they gave the opposite sign; according to the theoretical models, the periods should be *in*creasing.

In the 1991 paper describing their WET observations of this star, Winget and his coworkers found the rate of period change to be about 8 s per 10,000 years, similar to their 1985 result and still with the opposite sign from the theoretical models. This continuing disagreement prompted J. E. S. Costa, S. O. Kepler, and Winget to attack the problem again in 1999.[17] This time, with new data—now extending over the entire decade from 1983 to 1993—and using new methods of analysis, they found the period of this pulsation mode to be increasing by about 43 s per 10,000 years, giving a characteristic evolutionary timescale of 120,000 years. This result has the same sign as that predicted by the theoretical calculations, but the indicated timescale for changes in the mode is about ten times *faster* than the change given by the theoretical models.

Potential explanations for this difference that were advanced by the authors include the possibility that PG 1159 may have a more oxygen-rich core than the models indicate, which would lead to faster cooling; that neutrino emission from the degenerate core may be stronger, leading to faster evolution in earlier phases;

[15]Winget, D. E., *et al.* 1985, *Astrophys. J.*, **292**, 606, "A Measurement of Secular Evolution in the Pre-White Dwarf Star PG 1159-035."

[16]Kawaler (1986), *op. cit.*

[17]Costa, J. A. S., Kepler, S. O., and Winget, D. E. 1999, *Astrophys. J.*, **522**, 973, "Direct Measurement of a Secular Pulsation Period Change in the Pulsating Hot White Dwarf PG 1159-035."

or that the mass of the He-rich surface layer may be different than assumed in the theoretical models. As of this writing, this discrepancy is still unresolved.

Observations using the Whole Earth Telescope have also enabled astronomers to obtain physical data for the DBV stars, which lie farther down the white dwarf cooling track than the PG 1159 variables. In 1994, Winget and his colleagues carried out an extensive asteroseismological analysis of the prototypical DBV, GD 358, based on 154 h of WET data obtained in May 1990.[18] They found 180 significant peaks in the power spectrum, and they assumed all the triplets to be $l=1$ modes. Their detailed analysis yielded the stellar mass $M=(0.61\pm0.03)\,M_{Sun}$, and they determined the mass of the surface helium layer to be $(2.0\pm1.0)\times10^{-6}$ times the mass of the star (surprisingly low compared to the value of about $10^{-2}\,M_{Sun}$ anticipated). They also found the stellar luminosity to be $(0.050\pm0.012)\,L_{Sun}$ from which they obtained a seismic distance $d=(42\pm3)$ parsecs. In addition, they found strong evidence for differential rotation in this star, with the surface layers rotating 1.8 times faster than the core, and they were able to determine a value of $(1{,}300\pm300)$ gauss for the magnetic-field strength in the star.

All seemed well until 1996, when further observations happened to catch GD 358 undergoing a completely unexpected and still mysterious event. The amplitudes and frequency structure of the power spectrum underwent a dramatic change during a matter of hours. The astronomers termed this a "*sforzando*" event.[19] Of course, this prompted continuing investigations of the star. In 2009, Judi Provencal and a large team of colleagues reported the results of their investigations of GD 358 based on 436 h of WET observations in 2006.[20] Among other findings, they showed that the frequency multiplets cannot all be $l=1$ modes in slow rotation and that the 1996 *sforzando* event coincided with a long-term change in the multiplet splittings. They concluded that convection in the surface layers of this star diminished for some unknown

---

[18] Winget, D. E., *et al.* 1994, *Astrophys. J.*, **430**, 839, "Whole Earth Telescope Observations of the DBV White Dwarf GD 358."

[19] Provencal, J., *et al.* 2009, *Astrophys. J.*, **693**, 564, "2006 Whole Earth Telescope Observations of GD 358: A New Look at the Prototype DBV."

[20] *Ibid.*

reason during the 1996 event, and they speculated that perhaps the changes might be explained by changes in the internal structure of the magnetic field. Clearly, more remains to be learned about this white dwarf.

Astronomers have also conducted detailed asteroseismological analyses of other degenerate stars, including some ten DAVs, seven DBVs, and five DOVs/PNNVs. The results have provided values for white dwarf rotation periods, masses, distances, and masses for the surface hydrogen and helium layers. Not all of these physical quantities can be obtained for every star, however. The measured rotation periods range from a few hours to a few days, with most white dwarfs rotating with a period P_{rot} of about 1 day. The range of white dwarf masses determined from asteroseismology is essentially the same as that found by other methods, with a mean mass of about 0.6 M_{Sun}. Asteroseismological distance determinations range from 23 pc (for L19-2) to 670 pc (for the binary PG 1159 star PG 2131+066).

Asteroseismological determinations are generally consistent with theoretical expectations for the masses of the surface H layers (about 10^{-4} M_{Sun} for the DAV stars) and of the surface He layers (about 10^{-2} M_{Sun}), although there are some notable exceptions. The mass of the surface H layer of the DAV white dwarf GD 154 appears to be only about 10^{-10} M_{Sun}, for example, while the mass of the surface He layer of the DBV white dwarf EC 20058+5234 seems to be less than 10^{-6} M_{Sun}. The reasons for these discrepancies are not currently understood, though one possibility may be that the pulsation modes used to determine these values may not have been correctly identified.

19. Magnetic Personalities

I was working in my office at the University of Rochester one afternoon in 1969 when the telephone rang. When I answered, the caller identified himself as Jim Kemp, a physicist from the University of Oregon. He had recently shown that the continuum emission from a star with a strong magnetic field should be circularly polarized, and he had confirmed the effect with a simple laboratory experiment. Now he was interested in looking for stars that might actually display this phenomenon, and white dwarfs seemed likely candidates to possess strong enough magnetic fields to make this possible.

I no longer recall what I may have suggested to him, but not long afterward he and his colleagues reported the detection of circularly polarized radiation from the white dwarf Grw+70° 8247.[1] The unique spectrum of this star includes several then-unexplained band-like absorption features at unusual wavelengths. Indeed, this very peculiarity was the reason Jesse Greenstein had assigned this one star its own spectral class.

When Kemp applied his theory to Grw+70° 8247, he obtained an estimate of 12 million gauss (=12 megagauss = 12 MG) for the strength of its magnetic field.[2] Even the field strength of the strongest permanent magnets on Earth is more than a thousand times smaller than the one Kemp obtained for this white dwarf! We now know that the peculiar spectral features seen in this star are actually due to Zeeman splitting of the Balmer lines of hydrogen in a magnetic field stronger than 300 MG—some 25 times stronger

[1] Kemp, James C., *et al.* 1970, *Astrophys. J.*, **161**, L77, "Discovery of Circularly Polarized Light from a White Dwarf."

[2] For comparison, the strength of the field at Earth's magnetic poles is only about two-thirds of one gauss, while the field strength may exceed 10,000 gauss (=10 kilogauss = 10 kG) at the poles of a strong permanent magnet.

H.M. Van Horn, *Unlocking the Secrets of White Dwarf Stars*, Astronomers' Universe, DOI 10.1007/978-3-319-09369-7_19

than Kemp's original estimate and 30,000 times stronger than the strongest terrestrial magnets.

The successful discovery of circular polarization in Grw+70° 8247 spurred additional searches for magnetic white dwarfs, using measurements of circular polarization in the continuum as well as Zeeman splitting of the observed spectral lines. Additional members of the magnetic class were soon found.[3] Significant non-zero circular polarization has since been measured in about half the magnetic white dwarfs, and linear polarization has been observed in about a dozen. The magnetic field strengths, inferred either from polarization measurements or from the observed Zeeman splittings of line transitions, range from about 40 kG to enormous values, certainly in excess of 500 MG and perhaps as large as 1,000 MG (for PG 1031+234).

To interpret the spectra of magnetic white dwarfs, it is necessary to understand how the magnetic field affects the internal energy levels of atoms and ions and thus how it affects the spectra.

In general, the energy levels of an atom or ion are characterized by three quantum numbers, traditionally labeled n, l, and m, and transitions between different levels produce absorption or emission lines in spectra. In the absence of a magnetic field, these levels are degenerate with respect to the quantum number m; that is, they all have the same energy. A magnetic field removes this degeneracy, giving the different sublevels different energies. The amount of splitting depends on the strength of the field. For fields greater than a few megagauss, the magnetic field effects cannot be treated as small perturbations and instead require detailed numerical calculations.

Motivated by the discovery of the magnetic white dwarfs, several astrophysicists performed such computations beginning in the early 1970s. One of the most extensive of these efforts, carried out by Roy Garstang and S. B. Kemic at the Joint Institute for Laboratory Astrophysics (JILA) at the University of Colorado, obtained

[3] Angel, J. R. P., and Landstreet, J. D. 1970, *Astrophys. J.*, **160**, L147, "Magnetic Observations of White Dwarfs." See also Angel, J. R. P. 1978, *Ann. Revs. Astron. Astrophys.*, **16**, 487, "Magnetic White Dwarfs" and Angel, J. R. P., Borra, E. F., and Landstreet, J. D. 1981, *Astrophys. J. Suppl.*, **45**, 457, "The Magnetic Fields of White Dwarfs."

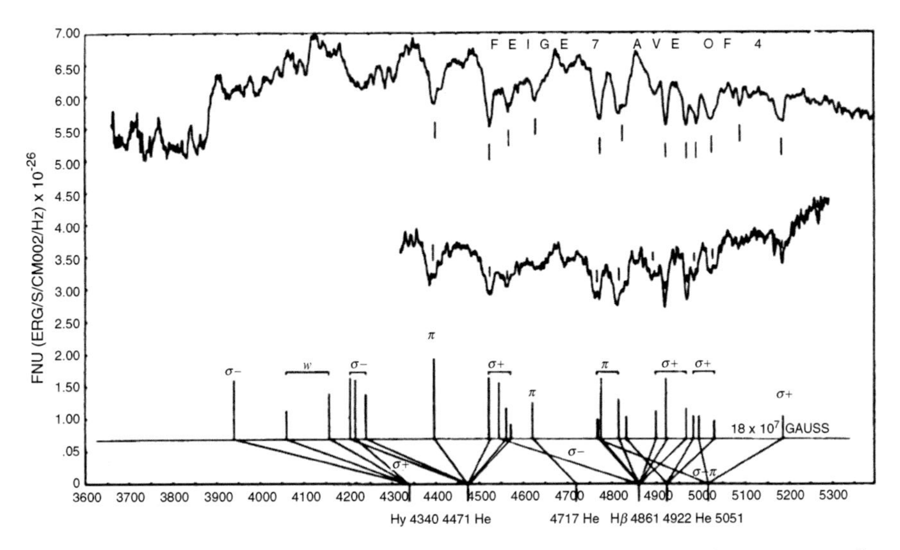

FIGURE 19.1 The spectrum of the magnetic white dwarf Feige 7. The upper curve shows the average of four image-tube scanner measurements obtained at Lick Observatory, while the lower curve is a digicon spectrum obtained at Steward Observatory. The Zeeman splittings of the Hβ line of hydrogen and several lines of neutral He, computed from Kemic's tables for a field of 18 MG, are shown in the lower portion of the figure. From Liebert *et al.* (1977). © AAS. Reproduced with permission

numerical solutions for the magnetic splitting in both H and He.[4] Their calculations extended up to 10 MG for atomic energy levels with principal quantum numbers up to $n=6$ and up to 100 MG for levels up to $n=3$.

An example of the application of such calculations is provided by the determination of the magnetic field strength in the DBAP3 white dwarf Feige 7.[5] As indicated by the DBA part of the spectral type, the atmosphere of this star contains both H and He, while the "P" denotes that the emission is polarized, and the "3" provides a rough indication of the effective temperature. The spectrum of this star is shown in the upper part of Figure 19.1, while the Zeeman splittings of the Hβ line of hydrogen and of several

[4] Garstang, R. H., and Kemic, S. B. 1974, *Astrophys. Space Sci.*, **31**, 103, "Hydrogen and Helium Spectra in Large Magnetic Fields." See also Kemic, S. B. 1974, *Astrophys. J.*, **193**, 213, "Hydrogen and Helium Features in Magnetic White Dwarfs."

[5] Liebert, J., *et al.* 1977, *Astrophys. J.*, **214**, 457, "Feige 7: A Hot, Rotating Magnetic White Dwarf."

lines of neutral He, all computed from Garstang and Kemic's calculations for a field of 18 MG, are shown in the lower part of the figure.

Because astronomers expect a range of field strengths to be present across the surface of a typical magnetic white dwarf, in general they also expect the magnetically split components of the spectral lines to be broadened into invisibility. However, a few components of each line exhibit either maxima or minima in wavelength as the field strength increases. If the range of surface field strengths in a magnetic white dwarf spans such an extremum in a line component, the line absorption is concentrated in that range of wavelengths rather than being broadened into invisibility, and an observable absorption feature is produced. For this reason, as Roger Angel pointed out in the late 1970s, observable features in high-field magnetic white dwarfs are expected to correspond to such extrema.

Over the next decade, a number of groups calculated the wavelengths and line strengths for several of the hydrogen lines that exhibit such extrema in high magnetic fields. These calculations made possible the discovery of extremely high magnetic field strengths in a number of white dwarfs. In particular, they permitted the determination of the field strength in Grw+70° 8247 by Jesse Greenstein[6] and others in the mid-1980s. Comparing the observed spectrum of this star with theoretical calculations of the hydrogen line shifts in fields up to 5,000 MG, they found that the locations of the extrema in the components of the Hα and Hβ lines at 320 MG match the observed features in the spectrum of this star very well.

About 10 % of all white dwarfs have been found to contain magnetic fields with strengths exceeding 1 MG.[7] Of these, about 60 % exhibit hydrogen features in their spectra, while another 15 % that have not yet been classified also are likely to have mainly H-dominated spectra. Some 18 % either display spectral features due to helium or have essentially continuous spectra and are probably He-dominated, while another 7 % exhibit strong,

---

[6] Greenstein, J. L. 1984, *Astrophys. J.*, **281**, L47, "The Identification of Hydrogen in Grw +70° 8247."

[7] Wickramasinghe, D. T., and Ferrario, Lilia 2000, *Publ. Astron. Soc. Pacific*, **112**, 873, "Magnetism in Isolated and Binary White Dwarfs."

distorted Swan bands of the C_2 molecule, which presumably indicate the presence of strong magnetic fields in these stars. These statistics are very similar to those of the non-magnetic white dwarfs. Magnetic white dwarfs are found throughout essentially the entire temperature range occupied by non-magnetic white dwarfs. In addition, several dozen magnetic white dwarfs are found in interacting cataclysmic variable systems such as AM Herculis or DQ Herculis (see Chap. 20).

In general, the axis of the field in a magnetic white dwarf is likely to be offset from its rotation axis, and in consequence many of these stars exhibit rotationally modulated variations in broadband polarimetry or spectrophotometry. The measured rotation periods of these stars range from 12 min (corresponding to a rotation velocity of about 80 km s^{-1}) to 18 days (corresponding to about 40 m s^{-1}). Rotation periods have not yet been measurable in some stars, however, and may indicate that their rotation periods exceed 100 years; Grw+70° 8247 is one of these. As we shall see below, information derived from rotation is important for quantitative modeling of the magnetic geometry of these stars.

To obtain detailed information from the spectrum of a magnetic white dwarf, it is necessary to compute a model for the stellar atmosphere. However, radiative transfer through the surface layers of a magnetic white dwarf is much more complicated than for a non-magnetic atmosphere. For example, one important effect of a strong magnetic field is to suppress convection. Even a field of only 0.1 MG is sufficient to accomplish this. Consequently, the thermal structure of an atmosphere that is sufficiently cool to be convective in the absence of a magnetic field will be quite different when a strong field is present. In addition, because a magnetic field affects the energy levels of atoms and ions, it also affects the excitation and ionization equilibria and thus the equation of state of stellar matter. These effects are small, however, except for field strengths in excess of about 100 MG and temperatures less than about 10,000 K.

As is the case for non-magnetic atmospheres, the opacities needed to construct a detailed model with a strong magnetic field include free-free (magneto-bremsstrahlung), bound-free, and bound-bound (line) opacities. Expressions for the first of these are well-known, and calculations of line opacities in strong magnetic fields also have been carried out. In addition to thermal broadening (due to the motions of ions and electrons along the line of sight) and pressure broadening, strong magnetic fields also produce "line-shift broadening" and "field-spread broadening." Departures from earlier approximate calculations are relatively minor, however. In addition, the energy levels of the free electrons in a strong magnetic field correspond to circular orbits around the magnetic field lines—called "cyclotron orbits" or "Landau levels"—in the plane perpendicular to the direction of the magnetic field.

Although transitions between the Landau levels of the free electrons can produce very strong absorption at the cyclotron frequency, they generally do not produce significant spectral features, because the cyclotron resonance is very narrow. Detailed calculations of the bound-free absorption have also been carried out and exhibit strong resonances in each of the Landau continua.[8]

Further, the plasma in a magnetic star is anisotropic, and calculations of radiative transfer involve vector and tensor quantities, rather than simply scalar ones. The theory of radiative transfer must therefore be generalized to take into account these effects. This has been done by a number of scientists. In 1979, B. Martin and D. T. Wickramsinghe summarized a very general formulation,[9] which has since been used extensively for computations of model atmospheres of magnetic white dwarfs. In a paper in 1986, they also took into account the effect of strong magnetic fields in suppressing convection in magnetic stellar atmospheres.[10]

Except for that effect, Martin and Wickramasinghe assumed that the thermal and mechanical structure of the atmosphere of a white dwarf are not affected by the presence of a strong magnetic field. After computing a flux-constant model atmosphere using conventional numerical methods, they started at the deepest point in the model, solved the transfer problem for polarized radiation, and moved outward step-by-step, until they obtained the solution for the spectrum of emergent intensity and linear and optical polarization.

---

[8] Merani, N., Main, J., and Wunner, G. 1995, *Astron. & Astrophys.*, **298**, 193, "Balmer and Paschen Bound-Free Opacities for Hydrogen in Strong White Dwarf Magnetic Fields."

[9] Martin, B., and Wickramasinghe, D. T. 1979, *Mon. Not. Roy. Astron. Soc.*, **189**, 883, "Solutions for Radiative Transfer in Magnetic Atmospheres."

[10] Wickramasinghe, D. T., and Martin, B. 1986, *Mon. Not. Roy. Astron. Soc.*, **223**. 323, "Magnetic Blanketing in White Dwarfs."

Because magnetic field lines intersect the photosphere at different angles and with different strengths over the surface of a magnetic white dwarf, the Zeeman splitting and polarization differ from point to point. Observations cannot resolve the disk of a white dwarf, however; instead they yield an average over the stellar surface. Obliquely rotating stars, which present an observer with varying views of the magnetic field structure, allow the most information to be determined about the field geometries. Researchers have attempted to model the spectra and spectropolarimetry of magnetic white dwarf stars with dipolar magnetic geometries, by analogy with the magnetic Main Sequence stars. However, detailed modeling of the Zeeman line components, spectropolarimetric energy distributions, and polarimetric light curves has generally shown that a simple, centered dipole is not sufficient to fit the observed properties of the star. Researchers have had to add complications to the dipole geometry, such as an offset of the dipole parallel or perpendicular to the rotation axis, or the introduction of a quadrupole component, or a high-field "spot." Unfortunately, the complications added to the field geometry for a given star seldom yield a unique solution to the observational data.

In their pioneering 1978 work on the magnetic field structure of the white dwarf BPM 25114,[11] Martin and Wickramasinghe first solved the equations of transfer for polarized radiation for several different inclination angles between the magnetic axis and the line of sight, assuming the field distribution to be given by a centered dipole. Then they computed the spectrum and polarization by summing up the results over the entire disk of the star. They and others subsequently extended this type of calculation to other magnetic white dwarfs, including off-center dipole fields as well as centered dipoles.

In the mid 1980s, Gary Schmidt at the University of Arizona and his collaborators investigated the magnetic field structure of the degenerate star PG 1031+234, finding very intense fields in the range from 200 to 500 MG.[12] Beginning with a trial field geometry at a given rotational phase of this star, they computed the

[11] Martin, B., and Wickramasinghe, D. T. 1978, *Mon. Not. Roy. Astron. Soc.*, **183**. 533, "A Dipole Model for Magnetic White Dwarf BPM 25114."

[12] Schmidt, G. D., *et al.* 1986, *Astrophys. J.*, **309**, 218, "The New Magnetic White Dwarf PG 1034+234: Polarization and Field Structures at More than 500 Million Gauss."

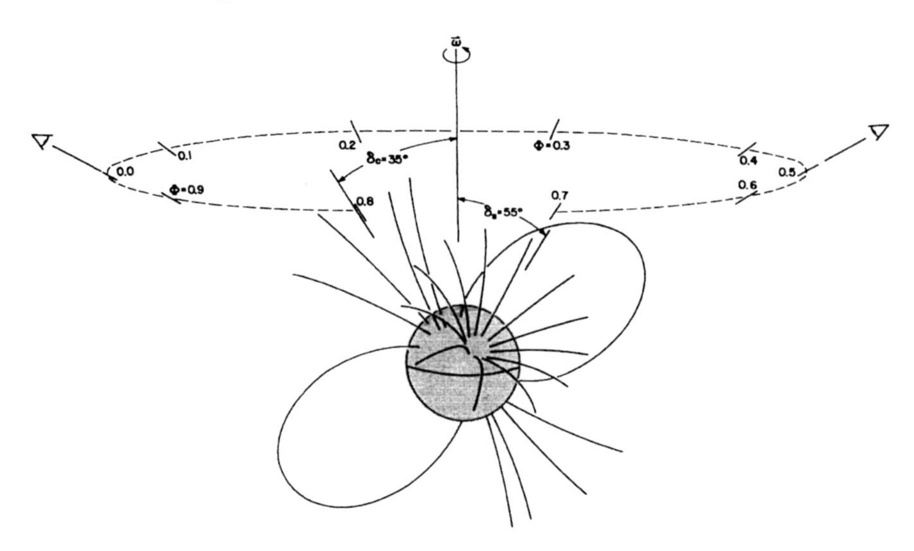

Figure 19.2 A sketch of the magnetic field distribution for a model for the strongly magnetic white dwarf PG 1031+234. The relative orientations of the axis of the main magnetic dipole, the axis of rotation, and the line of sight are indicated, and the rotational phases Φ are labeled. Note the intense magnetic spot, which faces the observer near $\Phi \approx 75°$. From Schmidt *et al.* 1986. © AAS. Reproduced with permission

spectrum and polarization for the observable hemisphere in essentially the same way as Martin and Wickramasinghe had done. Then Schmidt and collaborators repeated this procedure for several different rotational phases. Their final model consisted of a primary, centered, oblique dipole field of strength somewhere in the range from 200 to 500 MG, together with an intense magnetic spot with a field strength of about 700 MG, some two and a half times larger than the main dipole field. Figure 19.2 shows the geometry of this model.

In 1988, Dayal T. Wickramasinghe and his young colleague, Lilia Ferrario—now herself a professor in the Department of Mathematics at the Australian National University—performed similar modeling for the spectrum and polarization of the strongly magnetic white dwarf Grw+70° 8247.[13] They found good agreement with the observations for a polar field of strength 320 MG, inclined at an angle of 30° to the line of sight. Subsequent work by others essentially confirmed these results.

[13] Wickramasinghe, D. T., and Ferrario, Lilia 1988, *Astrophys. J.*, **327**, 222, "A Centered Dipole Model for the High Magnetic Field White Dwarf Grw +70° 8247."

The masses of the magnetic white dwarfs range from 0.32 to 1.35 M_{Sun}, with a mean mass of about 0.93 M_{Sun}, appreciably greater than the mean of about 0.6 M_{Sun}[14] for non-magnetic stars. Although it is difficult to determine masses for white dwarfs in which the spectral features are distorted by strong magnetic fields, several other lines of evidence also suggest greater masses for magnetic white dwarfs. This is reasonable, if the magnetic white dwarfs do indeed originate from Main Sequence stars with higher-than-average masses, as has been suggested.

How were the magnetic white dwarfs formed? Observational evidence indicates that most of those with field strengths in excess of 50 MG originated from chemically peculiar and strongly magnetic A and B stars (Ap/Bp stars) on the Main Sequence.[15] The fields in these white dwarfs are surviving "fossil" remnants of the magnetic fields observed in the Ap/Bp stars, which have measured magnetic field strengths in the range from 0.2 to 25 kG. Although the field slowly decays over time because of ohmic dissipation in the highly ionized and thus highly conducting stellar interior, calculations yield very long timescales for this—more than 10 Gyr for the Sun[16] and similarly long for a 0.6 M_{Sun} white dwarf.

Because the field decays so slowly, the field lines are effectively "frozen" into the stellar matter, so that the matter and field lines must move with each other in directions locally *perpendicular* to the field lines (although matter can flow freely *parallel* to the field lines). Drawings like that in Figure 19.2 illustrate the strength and direction of the invisible magnetic field using fictitious "magnetic field lines." Where they bunch close together—as at the magnetic poles or the strongly magnetic "spot" in this illustration—the field is stronger, and the tangent to a field line at a given point shows the direction a compass would point if placed there.

Because the magnetic field lines are "frozen" into the star, conservation of magnetic flux—which is proportional to the field strength multiplied by the square of the stellar radius—accounts

[14] See Chap. 15.

[15] Wickramasinghe, D. T., and Ferrario, L. 2005, *Mon. Not. Roy. Astron. Soc.*, **356**, 1576, "The Origin of the Magnetic Fields in White Dwarfs."

[16] Cowling, T. G. 1945, *Mon. Not. Roy. Astron. Soc.*, **105**, 166, "On the Sun's General Magnetic Field."

for the increase in the field as a star becomes a white dwarf. The radius of a white dwarf is only about one hundredth as large as the inner part of the Ap star from which it originated. Since magnetic flux is conserved, the field strength in a white dwarf must be about 10,000 times larger than that of the Ap/Bp star from which it originated in order to compensate for the shrinkage in radius as the star becomes a white dwarf. Consequently, field strengths of 0.2–25 kG on the Main Sequence produce fields of strength 2–250 MG in the white dwarf phase, comparable to the range of white dwarf fields that is observed. This hypothesis is also consistent with the preponderance of basically dipolar geometries observed in the white dwarfs and with other observational evidence.

However, some magnetic white dwarfs must have quite different origins. For example, the star EUVE J0317-853—the most massive of the magnetic white dwarfs yet discovered, with a mass of about 1.35 M_{Sun}—has a field strength of about 800 MG and a rotation period of only 726 s. In 1997, Lilia Ferrario and her collaborators proposed that this strongly magnetic degenerate originated from the merger of two white dwarfs of lower mass, at least one of which initially contained a strong field itself.[17] If these stars were originally in a very close binary orbit, the star with lower mass would have been disrupted by the strong tidal forces when it came too close to the more massive star. Matter from the disrupted star then formed a disk around the more massive white dwarf, which spun up to its current rate of rapid rotation as it accreted the remnants of its original companion from the disk. The strong shear flows accompanying this process would also have amplified the magnetic field, producing the strong field it now possesses.

Ferrario and Wickramsinghe in 2005 elaborated on the merger scenario and added a third possible origin for magnetic white dwarfs.[18] They pointed out that some of these stars have masses as low as 0.4 M_{Sun} and cannot have been produced through the evolution of single stars: stars that are initially less massive than about

[17] Ferrario, Lilia, *et al.* 1997, *Mon. Not. Roy. Astron. Soc.*, **292**, 205, "EUVE J0317-855: A Rapidly Rotating, High-Field Magnetic White Dwarf."

[18] Ferrario, L., and Wickramasinghe, D. T. 2005, *Mon. Not. Roy. Astron. Soc.*, **356**, 615, "Magnetic Fields and Rotation in White Dwarfs and Neutron Stars."

$0.8\,M_{Sun}$ burn hydrogen so slowly that they never leave the Main Sequence within the age of the Universe. Because white dwarfs with lower masses than this are known to exist, however, there must be some other pathway that allows this to happen. As we shall see in the next chapter, such pathways do exist; they involve the evolution of stars in binary systems that are sufficiently close to enable mass exchange between, and mass loss from, the two stars. Ferrario and Wickramasinghe pointed out that such binary star evolution can spin up white dwarfs to rotation periods of the order of hours and suggested that the magnetic white dwarf PG 1031+234, with a rotation period of 3 h and 24 min, may have originated in this way.[19]

Yet another possibility is that the strong fields seen in some magnetic white dwarfs may be produced during the final, asymptotic giant branch (AGB) phase of pre-white dwarf evolution.[20] Calculations show that the He-burning shell sources in the AGB stars develop a thermal pulse instability. Each thermal pulse results in the formation of an extended convection zone that spans most of the outer layers of the star. Theory and observation agree that the outer layers of these red supergiants rotate very slowly, while the dense inner cores are expected to rotate quite rapidly. Under these conditions, a magnetohydrodynamic dynamo like that in our Sun may operate near the base of the deep convective envelope during the thermally pulsing phase of AGB evolution. Time-dependent model calculations suggest that field strengths of order 10 kG may be generated in such a dynamo. During the final contraction of the AGB core to the white dwarf stage, field strengths up to about 400 MG might be produced in this way.

In summary, about 10 % of all white dwarfs have now been found to have enormously strong magnetic fields, with field strengths of many hundreds of megagauss. Such fields are strong enough to produce extreme distortions of the internal quantum states of atoms, ions, and molecules and thus of the optical spectra and material properties of the surface layers of these stars. Except for the strong fields, the magnetic white dwarfs appear to share the

[19] *Ibid.*

[20] Blackman, Eric G., *et al.* 2001, *Nature,* **409**, 485, "Dynamos in Asymptotic Giant Branch Stars as the Origin of Magnetic Fields Shaping Planetary Nebulae."

same characteristics as their non-magnetic "cousins." It seems likely that most (but not all!) magnetic white dwarfs originated from Main Sequence Ap and Bp stars, which also display stronger than average magnetic fields. Whether this postulated evolutionary connection will eventually enable astronomers to learn more about the processes by which a star becomes a white dwarf, however, remains today an open questions.

20. Odd Couples

White dwarf stars harbor yet more surprises. Consider the Sirius system, for example. Sirius A, with a mass of $2\,M_{\text{Sun}}$, is an ordinary, hydrogen-burning Main Sequence star. Sirius B, with a mass of $1\,M_{\text{Sun}}$, is a white dwarf. It has already completed hydrogen- and helium-burning and has no remaining nuclear energy sources. However, stellar evolution calculations tell us that stars with higher masses consume their nuclear fuels faster than those with lower masses. Thus, Sirius B must once have been more massive than Sirius A, and sometime during the course of stellar evolution it must have lost much of its original mass. Astronomers think that most or all of this mass was ejected completely from the Sirius system.[1]

As this example illustrates, stellar evolution in a binary system is a much richer process than is the evolution of a single star. Depending on the initial masses of the two stars and their orbital separation, many different outcomes are possible. The effort to understand the evolution of such systems is important, for about half—and perhaps more than two-thirds—of all stars in the sky are members of binary or multiple star systems. In their initial Main Sequence phases, the two components of stellar binaries range in mass from about $0.1\,M_{\text{Sun}}$—approximately the lower limit in mass at which a body can ignite hydrogen burning and become a real star—up to more than $60\,M_{\text{Sun}}$. The semi-major axes of their orbits range from just a few times the radius of the Sun up to more than 30 AU.

To understand how stellar evolution in a binary system may form a white dwarf, it is helpful to refer to the so-called "Roche model,"[2] which considers two point masses in circular orbits

[1] Holberg (2007), p. 215.

[2] See, for example, Kopal (1959).

H.M. Van Horn, *Unlocking the Secrets of White Dwarf Stars*,
Astronomers' Universe, DOI 10.1007/978-3-319-09369-7_20

around each other.[3] In a reference frame co-rotating with the two objects, the surfaces of equal (gravitational plus "centrifugal") potential energy are nearly spherical close to either mass, while farther away they become increasingly egg-shaped, with the narrower end of each surface pointing in the direction of the companion. Up to a certain distance, the shape becomes more extreme farther away from the star, and it comes to a point at a unique, teardrop-shaped surface called a "Roche lobe." Each mass is enclosed within its own Roche lobe, with the pointed ends of the teardrop shapes in contact with each other. Within the Roche lobe surrounding one component of a binary star system, matter can be considered as belonging to that star. The point at which the Roche lobes for the two stars touch is called the "inner Lagrange point." At still larger distances, the equipotentials again become nearly spherical surfaces centered on the center of mass of the binary system.

At several different stages in the evolution of a star, its radius expands, in some cases by very large factors. These phases include evolution off the Main Sequence; rapid evolution up the red giant branch (RGB) to the ignition of helium burning; and rapid evolution up the asymptotic giant branch (AGB), as the carbon/oxygen core becomes increasingly degenerate just before the star ejects its outer layers into space to form an expanding shell of ionized gas called a planetary nebula. In a binary system, as a star's outer layers expand along the RGB or AGB, the stellar surface takes on the shapes of the equipotential surfaces. The outer layers become increasingly egg-shaped as the star expands or the orbit contracts, and when the surface of the star fills its Roche lobe, matter can escape through the inner Lagrangian point into the Roche lobe surrounding its companion. This transfer of mass from one star to the other affects the subsequent evolution of both stars, as well as the evolution of the binary orbit, as first pointed out in a pioneering work by Rudolf Kippenhahn and Alfred ("Arlie") Weigert in 1967.[4]

---

[3] This model was first developed by French astronomer and mathematician Èdouard A. Roche (1820–1883) and has since been used extensively in studies of close binary systems.

[4] Kippenhahn, R., and Weigert, A. 1967, *Zeitschrift für Astrophysik*, **65**, 251, "Entwicklung in engen Doppelsternsystemen. I. Massenaustausch vor und nach Beendigung des zentralen Wasserstoff-Brennens."

In 1989, Ronald F. Webbink at the University of Illinois published a particularly readable summary of the results of stellar evolution in a close binary system,[5] and we largely follow his discussion in the following. Consider a close pair of Main Sequence stars with comparable masses somewhere in the range from about 2 to 8 M_{Sun}. If the primary—the more massive of the two stars, which is also the first to finish hydrogen burning and leave the Main Sequence—does not fill its Roche lobe until just before it ignites nuclear burning of He at its center, the rate of expansion of the stellar envelope as the star climbs the red giant branch is so fast that mass lost from the primary overwhelms the Roche lobe of its companion and forms a common envelope surrounding both stars—a viscous medium within which the binary system evolves. The orbital motion of the two stars acts like an "eggbeater" that drains energy and angular momentum from the orbit and deposits it in the common envelope, driving it away. The dissipation of energy and angular momentum also causes the two stars to spiral gradually toward each other, so that the final orbital separation may be much smaller than the orbital separation of the original pair of Main Sequence stars. Cataclysmic variables (see below) and close binary central stars of planetary nebulae may be formed in this way.

The common envelope phase continues until the amount of H-rich matter left on the surface of the primary decreases to about the thickness of the H-burning shell. The remnant then contracts until helium burning is ignited at its center. According to calculations by Iben,[6] if the initial Main Sequence mass of the primary is 10 M_{Sun}, the mass of the core-He-burning star that remains after mass transfer is 2 M_{Sun}, while if the initial mass is 5 M_{Sun}, the remnant mass is about 0.77 M_{Sun}.

If the initial Main Sequence mass of the primary star is less than about 2.3 M_{Sun}, evolution in a close binary system converts it into a helium white dwarf with a mass less than about 0.5 M_{Sun}. In this case, the low-mass remnant of the original primary star that survives the common envelope event never gets hot enough to ignite core He-burning.

Conversely, if the initial mass of the primary lies in the range from about 2 to 10 M_{Sun}, close binary evolution converts it into a

[5] Webbink, R. F. 1989, *American Scientist*, **77**, 248, "Cataclysmic Variable Stars."
[6] Iben, I., Jr. 1991, *Astrophys. J. Suppl.*, **76**, 55, "Single and Binary Star Evolution."

CO white dwarf with mass in the range from about 0.25 to 1.1 M_{Sun}. In these cases, core He-burning converts the central regions of the star into a mixture of carbon and oxygen (CO), and the star expands again to fill its Roche lobe at the completion of central He burning. It continues to lose mass, first ridding itself of the residual amount of H-rich matter and then transferring the He-rich material thus exposed at the stellar surface, until the He layer has been reduced below some (small) critical value. In this way, a 2 M_{Sun} remnant from the first mass-exchange event becomes a 1.05 M_{Sun} CO white dwarf, while a 0.77 M_{Sun} remnant becomes a 0.75 M_{Sun} CO white dwarf.

If the initial mass exceeds about 9–11 M_{Sun}, close binary evolution ultimately converts it into a white dwarf composed of some mixture of oxygen and neon (ONe), with mass in the range from about 1.1 to 1.4 M_{Sun}. The ONe composition results from He-burning at the higher temperatures and densities in these more massive remnants of common envelope evolution.

The interesting results of stellar evolution in a binary system do not end with the formation of a white dwarf, however. For example, close binary systems that contain a white dwarf and a low-mass Main Sequence star may become "cataclysmic variables" (CVs) when the companion star finishes its own Main Sequence phase of H-burning, expands, and overflows onto the white dwarf. These systems are relatively inconspicuous in quiescence, but the transfer of matter back onto the white dwarf can produce outbursts of energy. These rapid increases in light output may happen on a timescale as short as a single day, while the subsequent decline in luminosity may take weeks or months. Eruptions may repeat in days to years or even longer.

Other close binary systems, which may become other kinds of CVs, may contain a white dwarf and a red giant or, in rare cases, another white dwarf. Their orbital periods range up to hundreds of days for systems that contain red giant companions, while systems containing a white dwarf and a low-mass companion extend from about 1 to 15 days, and those containing two white dwarfs may have orbital periods as short as about 20 min.

More than 1,600 cataclysmic variable systems are known.[7] They come in several different types, including dwarf novae,

[7] Downes, R., *et al.* 2001, *Publ. Astron. Soc. Pacific*, **113**, 764, "A Catalog and Atlas of Cataclysmic Variables."

helium CVs, strongly magnetic systems, and classical novae. These are briefly described below, following in part the summaries provided by Carl Hansen and Steve Kawaler.[8]

Some 300 dwarf novae are known.[9] They erupt repetitively, though not with regular intervals, and tens to hundreds of days may elapse between outbursts. The longer the duration of an outburst, the longer the time between outbursts. There is little or no mass lost from the system during the event. At peak, a dwarf nova is about 10 times as bright as the Sun. The total energy released during an outburst is some 10^{38}–10^{39} ergs, equivalent to the total power output of the Sun for about 1–10 days.

In a cataclysmic variable binary system, the rotation of the stars in their mutual orbit causes the mass transferred through the inner Lagrange point to spiral in toward the companion. If the transfer rate is not too large, this stream of matter forms a flattened structure called an "accretion disk"—a thin disk of gaseous matter that orbits in nearly circular annular rings around a central star. Symmetric about its mid-plane, the disk is contained entirely within the star's Roche lobe. Matter is supported in the direction perpendicular to the disk by the thermal pressure of the gas, which in turn is maintained by energy generated through viscous dissipation of the shear between adjacent annular rings. At the same time, the disk is confined to the region about the mid-plane by the component of the gravity produced by the central star that is perpendicular to the plane of the disk.

The thickness of an annulus of the disk is much smaller than the radius of the annulus, and the disk is thicker at its outer edge than it is nearer to the central star. Viscous dissipation of energy within the disk allows matter transferred from the primary star to flow through the disk and be accreted onto the surface of the companion. Substantial energy may be released in the process, in some cases causing a considerable increase in the luminosity of the system. There is clear observational evidence for a stream of matter directed from the cool star into the Roche lobe surrounding the white dwarf. There, the stream of infalling matter collides with the outer edge of the accretion disk to produce a "hot spot" (see Figure 20.1).

[8] Hansen and Kawaler (1994), pp. 83 ff.

[9] *Ibid.*, p. 84.

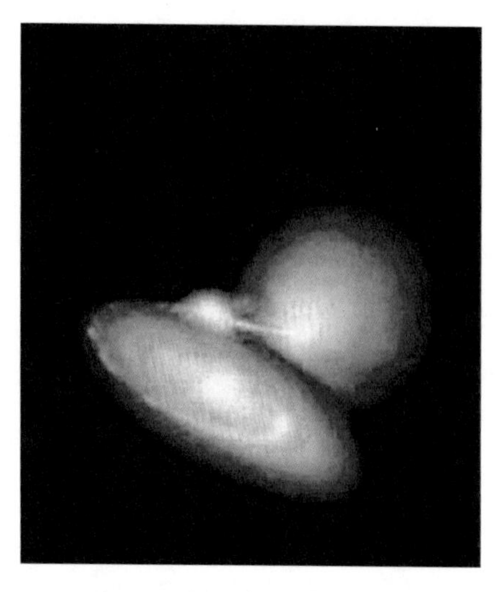

Figure 20.1 Artist's concept of a dwarf nova system. The red Main Sequence star in the background fills its Roche lobe and is in the process of transferring mass to its white-dwarf companion. The stream of matter impacts the outer edge of the accretion disk surrounding the white dwarf, producing a "hot spot." Matter flows through the accretion disk and onto the white dwarf, releasing energy as it is accreted onto the white dwarf. Public domain image by Dana Berry, courtesy of NASA

Astrophysicists think that the cause of a dwarf nova outburst is a thermal instability in the accretion disk itself. This instability depends upon the rate of mass transfer and the rate of energy dissipation in the disk, as originally pointed out in 1974 by Japanese astrophysicist Yoji Osaki[10] and subsequently explained in more detail by others. If the mass is transferred sufficiently slowly, the disk remains relatively cool and partially transparent, and the energy liberated through viscous dissipation can be carried to the disk surface by convection and radiated away. Matter can thus flow steadily inward through the disk until it is accreted onto the surface of the white dwarf.

However, there is no reason why mass transfer must necessarily proceed so slowly. If in fact matter is transferred from the donor star at a rate faster than it can be accommodated by viscous energy dissipation and convective transfer, matter will accumulate in the disk. When this occurs, the build-up of mass in

[10] Osaki, Y. 1974, *Pub. Astron. Soc. Japan*, **26**, 429, "An Accretion Model for the Outbursts of U Geminorum."

the annular rings causes the density and temperature at the disk mid-plane to rise. At some stage, the disk becomes hot enough to ionize hydrogen, causing a transition to efficient radiative transfer of energy vertically to the disk surface, and a dramatic increase in the viscous dissipation of energy. This produces a large increase in the energy emitted from the surface of the disk and a simultaneous large increase in the inward rate of mass transfer, in effect rapidly dumping matter that had been stored in the disk onto the surface of the white dwarf. After the stored matter has been drained from the disk, it returns to its initial state. The continuing transfer of mass from the donor star then causes matter to start building up again in the disk, setting the stage for a repetition of the cycle.

There are several different types of dwarf novae. The "U Geminorum" systems are prototypical dwarf novae, which behave as just described. Another type, the "Z Camelopardalis" systems, have normal dwarf nova outbursts but in addition sometimes get "stuck" for days or months at an intermediate level of luminosity between that of outburst and that of quiescence. And a third type, the "SU Ursae Majoris" systems, undergoes additional "super outbursts," where the light output is much greater than in a normal outburst.

In contrast to the dwarf novae, the "AM Canum Venaticorum," or AM CVn stars, are hydrogen-deficient binaries that contain a degenerate CO white dwarf primary and a semi-degenerate, low-mass secondary. As described in 2010 by Jan-Erik Solheim,[11] the spectra of these systems show only He and heavy elements, because the low-mass secondary stars have lost their outer H-rich envelopes during a common envelope phase of binary evolution. In an AM CVn system, the low-mass secondary star loses He-rich matter to form an accretion disk around the white dwarf primary. There are 25 known AM CVn stars, 13 of which were found by the Sloan Digital Sky Survey.

In 1972, John Faulkner and his colleagues proposed a model for AM CVn in which a 0.041 M_{Sun} He white dwarf is the donor star.[12] This is an astonishingly small mass for a white dwarf!

[11] Solheim, J.-E. 2010, *Pub. Astron. Soc. Pacific*, **122**, 1133, "AMCVn Stars: Status and Challenges."

[12] Faulkner, J., Flannery, B. P., and Warner, B. 1972, *Astrophys. J.*, **175**, L79, "Ultrashort-Period Binaries. I. HZ 29 (=AM C Vn): A Double-White dwarf Semidetached Protocataclysmic Nova?"

Because the radius of a white dwarf increases as the mass decreases, and because the Roche lobes continue to shrink as the binary orbit shrinks, the very low-mass donor star continues to fill its Roche lobe and consequently to lose mass until all the matter it originally contained has been completely accreted onto its companion CO white dwarf.

The orbital periods of the AM CVn systems range from about 5 to 65 min. The orbital period of AM Cvn itself is 1,028.7 s, or about 17 min. Binary evolution is driven toward such ultra-short periods by gravitational radiation, as first proposed in 1962 by Robert P. Kraft,[13] with the orbit continuing to shrink as long as the emission of gravitational waves dominates. Gravitational radiation is one of the consequences of Einstein's general relativistic theory of gravity. Despite extensive and dedicated efforts to detect gravitational waves—including the Laser Interferometer Gravitational-wave Observatory (LIGO)—these efforts have not yet succeeded. There is, however, clear evidence that gravitational radiation exists, as pointed out by radio astronomers Russell Hulse and Joseph Taylor in their 1975 Nobel-Prize-winning work on the binary pulsar, PSR 1913+16.[14]

Another type of cataclysmic variable system includes a white dwarf with a strong magnetic field. Just as some isolated white dwarfs possess strong magnetic fields, so do some in close binary systems. If the white dwarf in a close binary does have a strong field, the nature of the mass flow between the two stars is substantially altered, and the formation of an accretion disk may even be prevented. A "polar" or "AM Herculis" binary contains a white dwarf with a magnetic field ranging in strength from about 10 to 230 MG. The strong field locks the two stars together so that they both rotate with the same period as the orbital period, always less than 3 h, with the magnetic pole of the white dwarf pointing at its companion. This configuration channels the accretion flow directly down onto the magnetic poles of the white dwarf. AM Herculis systems radiate circularly polarized light,

---

[13] Kraft, R. P. 1962, *Astrophys. J.*, **135**, 408, "Binary Stars Among Cataclysmic Variables. I. U Geminorum Stars (Dwarf Novae)."

[14] Hulse, R. A., and Taylor, J. H. 1975, *Astrophys. J.*, **195**, L51, "Discovery of a Pulsar in a Binary System."

which is typical of radiation produced in strong magnetic fields, and they may have high and low states of light output that may last for weeks to years. In contrast, "intermediate polars" or "DQ Herculis" binaries have lower magnetic fields. The two stars do not rotate synchronously in these systems, and the orbital periods are greater than 3 h.

The final type of cataclysmic variable we shall consider here is the classical nova. This is a close binary system in which a thermonuclear runaway occurs on the surface of a white dwarf that has accreted H-rich matter from its companion. This produces an outburst that on a timescale of weeks to months emits as much energy as the Sun produces in 1,000–10,000 years, and its explosive release ejects as much as $10^{-4} M_{\mathrm{Sun}}$ of mass at velocities of about 1,000 km s^{-1}. The outburst follows the gradual accretion of 10^{-5}–10^{-4} M_{Sun} of H-rich material from the donor star onto the surface of the white dwarf. By the time this much matter has accumulated, conditions at the base of the accreted layer reach temperatures and densities high enough to cause runaway thermonuclear burning of the accumulated hydrogen. Detailed hydrodynamic calculations are required to follow the development of a nova explosion, as is also the case for terrestrial thermonuclear devices. Sumner Starrfield at Arizona State University, Warren Sparks at Los Alamos National Laboratory, James Truran at the University of Illinois and their colleagues have been among the leaders in such nova calculations. Since the amount of mass necessary to trigger a nova outburst is similar to the amount of matter ejected during a nova explosion, it is still an open question whether an accreting white dwarf can continue to gain mass until it reaches the Chandrasekhar limit or whether the accreted mass is entirely expelled in the nova outburst.

There is also a population of double white dwarf binaries in the Milky Way Galaxy,[15] with orbital periods ranging from hours to days.[16] They are thought to be formed as follows:[17] After the first white dwarf has been formed, if the secondary star is massive

[15] As of 2005, 24 double white dwarf binaries have been discovered.

[16] Nelemans, G., *et al.* 2005, *Astron. & Astrophys.*, **440**, 1087, "Binaries Discovered by the SPY Project. IV. Five Single-Lined DA Double White Dwarfs."

[17] Webbink, R. F. 1984, *Astrophys. J.*, **277**, 355, "Double White Dwarfs as Progenitors of R Coronae Borealis Stars and Type I Supernovae."

enough to evolve off the Main Sequence by the time the Galaxy has reached its present age, a second common envelope event usually follows when it exhausts its nuclear fuel and expands to fill its own Roche lobe. Orbital shrinkage again occurs, and two white dwarf remnants emerge when the common envelope has been ejected.

The orbital separation of the two white dwarfs is likely to be very small. If it is less than about three times the radius of the Sun, gravitational radiation will cause the loss of orbital angular momentum, and the white dwarfs may again move into Roche lobe contact in less than about 10 billion years. If the first formed white dwarf has less mass than about 0.3 M_{Sun}, the ultimate result is two He white dwarfs that eventually merge. If the combined mass of the two white dwarfs is less than about 0.4–0.5 M_{Sun}, the lighter white dwarf may lose all its mass to the more massive white dwarf and disappear completely. The white dwarf 40 Eridani B, with a mass of 0.45 M_{Sun}, may have been formed this way, as suggested by Icko Iben and Alexander Tutukov in 1986.[18] If the two stars are He white dwarfs with comparable and sufficiently large masses, such a merger may produce a "hot subdwarf" star—an sdO or sdB star—that ultimately evolves into a CO white dwarf, as originally suggested by Ron Webbink in 1984.[19] If both are CO white dwarfs and if their combined mass exceeds the Chandrasekhar limit of about 1.4 M_{Sun}, the merger may produce a Type Ia supernova, as suggested in 1984 by Iben and Tutukov[20] and by Webbink[21]; we shall consider this possibility a bit further in the final chapter. If the more massive star is an ONe white dwarf, the merger may produce a neutron star but without an accompanying supernova explosion, as suggested by Iben 1986.[22]

[18] Iben, I., Jr., and Tutukov, A. 1986, *Astrophys. J.*, **311**, 753, "On the Number-Mass Distribution of Degenerate Dwarfs Produced by Interacting Binaries and Evidence for Mergers of Low-Mass Helium White Dwarfs."

[19] Webbink, R. F. 1984, *op. cit.*

[20] Iben, I, Jr., and Tutukov, A. 1984, *Astrophys. J. Suppl.*, **54**, 335, "Supernovae of Type I as End Products of the Evolution of Binaries with Components of Moderate Initial Mass (*M* not Greater than about 9 Solar Masses)."

[21] Webbink, R. F. 1984, *op. cit.*

[22] Iben, I., Jr. 1986, *Astrophys. J.*, **304**, 201, "On the Evolution of Binary Components Which First Fill Their Roche Lobes After the Exhaustion of Central Helium."

21. White Dwarfs and the Nature of the Milky Way Galaxy

By the end of the twentieth century, astronomers' understanding of the properties of white dwarfs had matured sufficiently that it had become possible to use these stars to make other astronomical measurements. But what kinds of measurements could they be used for? Because the white dwarf stars are so faint that they can only be detected out to distances relatively close to our Sun, at first it seemed unlikely that white dwarfs could have anything at all to tell us, *e.g.*, about the nature of the vastly larger Milky Way Galaxy in which we live. However, white dwarfs rather surprisingly *do* provide some information about our home Galaxy.

The Age of the Galactic Disk

Stretching across the night sky is an irregular band of light that has been called the Milky Way since ancient times. It is actually a cross-section of a vast disk of some 100 billion stars that we call the Milky Way Galaxy, as seen from our own edge-on perspective within it. At the center of the Galaxy is an enormous, flattened ball of stars a few kiloparsecs[1] in diameter and perhaps a kiloparsec thick, called the galactic bulge. From our vantage point on Earth, the bulge appears as a slightly wider expanse of the Milky Way in the constellation Sagittarius. And lurking at the very center of the bulge is a supermassive black hole containing as much mass as a million Suns or more. It is crammed into a space only a few times larger than the diameter of our Sun, and the force of gravity at its surface is so high that, like all black holes, not even light can escape.

[1] One kiloparsec, or 1 kpc, is a distance of 1,000 pc. It equals approximately 3,300 light-years.

H.M. Van Horn, *Unlocking the Secrets of White Dwarf Stars*, Astronomers' Universe, DOI 10.1007/978-3-319-09369-7_21

Long arms containing billions of stars—as well as gas and dust from which new stars are constantly being born—spiral outward from the central bulge to a distance of about 12 kPc from the galactic center. During the course of the twentieth century, astronomers discovered that our Solar System resides more than halfway out along one of these spiral arms. Surrounding the Galaxy is a huge, roughly spherical "halo" some 30 kpc in radius that contains a low density of very old stars, globular clusters of other ancient stars, and tenuous clouds and streams of gas falling in toward the disk from intergalactic space. Some of the groups of stars may be remnants of still older, smaller galaxies that merged with the Milky Way in the distant past.

Though we cannot see our own Galaxy from the outside, astronomers think that it resembles the Andromeda Galaxy pictured in Figure 21.1, which lies some 680 kPc away. The Milky

FIGURE 21.1 The Andromeda Galaxy. Credit: Bill Schoening, Vanessa Harvey/REU program/NOAO/AURA/NSF. Used with permission

Way, the Andromeda Galaxy, and a number of smaller galaxies are gravitationally bound to each other in a collection astronomers call the Local Group.

Since the Galaxy first came into existence, uncountable numbers of stars have formed, passed through their evolutionary life cycles, and died. The most massive stars die in spectacular supernova explosions, spewing back into space debris enriched in heavier chemical elements formed during various stages of nuclear burning. Subsequent generations of stars form from this debris.

Stars that are initially less massive than about $10\,M_{\text{Sun}}$—amounting to perhaps 90 % of all stars formed—shed enough mass during the course of their evolution to end as white dwarfs. The white dwarf stars consequently provide a "fossil record" of the history of star formation in the Galaxy, though to date, astronomers have only been able to extract the age of the galactic disk from this record. We will see below how they obtained this chronometric measurement from white dwarfs.

Thanks to Leon Mestel's insightful work in the early 1950s, astronomers knew that white dwarfs derive their luminosities by slowly radiating away their residual internal heat, and that the fainter they get, the more slowly they cool. Consequently, if one were to count the numbers of white dwarfs in given intervals of luminosity, one would expect to find more and more of the fainter and fainter stars. A graph showing this relationship is called the "white dwarf luminosity function," or WDLF.

In 1967, German astrophysicist Volker Weidemann made the first effort to test this prediction of Mestel's theory, using the meager amount of data then available.[2] Reasonably accurate values for the absolute magnitudes were then known for only a few hundred white dwarfs, and the statistical uncertainties in the resulting WDLF were too large to be able to tell much more than that the plot of observational data was more or less in agreement with Mestel's theory.

Over the decade and a half following Weidemann's original work, both the quantity and quality of the observational data used to construct the WDLF improved markedly. This was thanks in part to diligent work by Jim Liebert at the University of Arizona and to parallax studies—which provided distances and thus absolute

[2] Weidemann, V. 1967, *Zs. für Astrophysik*, **67**, 286, "Leuchtkraftfunktion und räumliche Dichte der Weißen Zwerge."

magnitudes—by Conard Dahn and David Monet at the U.S. Naval Observatory in Flagstaff, Arizona. Equally important were the detailed analyses of white dwarf atmospheres carried out by astrophysicists such as Harry Shipman at the University of Delaware, Detlev Koester and his colleagues in Germany, and François Wesemael, Pierre Bergeron, and others at the Université de Montréal.

By about 1980, astronomical detectors had become sensitive enough so that white dwarfs could be studied relatively routinely down to very faint magnitudes. Near the beginning of this decade, Jim Liebert and his colleagues began to suspect that there were fewer white dwarfs than were predicted by Mestel's theory below luminosities less than about 1/10,000th that of the Sun.[3] This discovery set off a flurry of speculation among theorists as to the cause of the shortfall. Could it be caused by some unusually rapid cooling mechanism that set in at low luminosities? Or was cooling instead delayed, for example by radioactive decay catalyzed by exotic fundamental particles such as hypothetical magnetic monopoles?

In the end, the most natural explanation proved to be that the shortfall results from the fact that these very old, faint white dwarfs are remnants of some of the very first stars formed in the disk of the Milky Way Galaxy. That is, no stars were formed earlier than this in the galactic disk, and the time required for them to finish their nuclear-burning phases of evolution, become white dwarfs, and cool down to the low luminosities at which they are observed today pre-dates the onset of star formation in the disk. In 1987, Don Winget and his colleagues used this approach to find the age of the galactic disk to be (9.3 ± 2.0) billion years.[4]

Liebert *et al.* continued their work through the end of the 1980s to improve the observational determination of the faint end of the WDLF.[5] In the early 1990s, Matt Wood, a former student of Don Winget's at Texas and by then a faculty member at the Florida Institute of Technology, similarly worked to improve theoretical models to refine the comparison with the observational WDLF.[6]

---

[3] Liebert, J., *et al.* 1979, *Astrophys. J.*, **233**, 226, "New Results From a Survey of Faint Proper-Motion Stars—A Probable Deficiency of Very Low-Luminosity Degenerates."

[4] Winget, D., *et al.* 1987, *Astrophys. J.*, **315**, L77, "An Independent Method for Determining the Age of the Universe."

[5] Liebert, J., Dahn, C., and Monet, D. G. 1988, *Astrophys. J.*, **332**, 891, "The Luminosity Function of White Dwarfs."

[6] Wood, M. A. 1992, *Astrophys. J.*, **386**, 539, "Constraints on the Age and Evolution of the Galaxy from the White Dwarf Luminosity Function."

Wood concluded that the largest uncertainties in the age of the galactic disk as determined by this method were due to uncertainties in the bolometric correction (the difference between the visual magnitude and the bolometric magnitude of a star, which is obtained from detailed model atmosphere calculations), uncertainties in the C/O ratio in the core of the star (which depend upon the nuclear reaction rates during He-burning), and on uncertainties in the masses of white dwarf surface layers.

By the end of the decade, both observations and theories had been greatly improved. In 1998, this enabled Sandy Leggett at the NASA Infrared Telescope Facility in Hawaii, María Teresa Ruiz of the Universidad de Chile, and Pierre Bergeron to compare the improved observational data for the faint end of the WDLF with improved theoretical models by Wood and others[7] to refine the age of the galactic disk to (8 ± 1.5) Gyr[8] (see Figure 21.2). The peak of their WDLF occurs at $M_{bol} = 14.75$, corresponding to $\log L/L_{Sun} = -4.0$

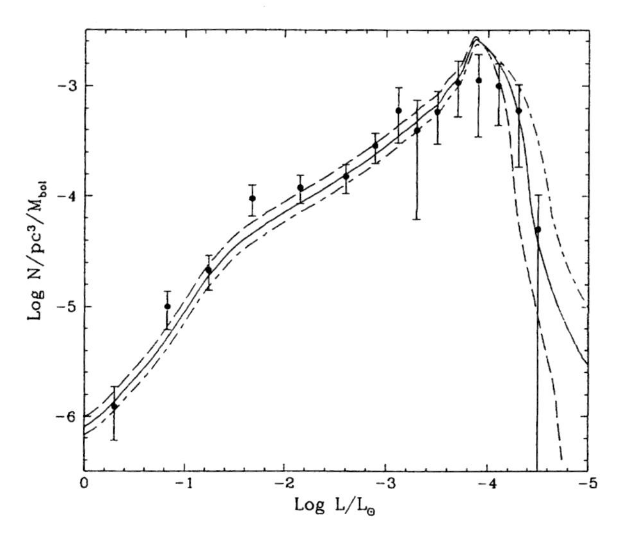

FIGURE 21.2 Comparison between the faint end of the WDLF obtained in 1998 by Sandy Leggett and her colleagues with theoretical models by Matt Wood to determine the age of the Galactic disk as approximately 8 Gyr. The solid curve corresponds to a disk age of 8 Gyr, while the dashed curve corresponds to 7 Gyr and the dot-dashed curve to 9 Gyr. From Leggett, Ruiz, and Bergeron 1998. © AAS. Reproduced with permission

[7] Leggett, S. K., Ruiz, M. T., and Bergeron, P. 1998, *Astrophys. J.*, **497**, 294, "The Cool White Dwarf Luminosity Function and the Age of the Galactic Disk."

[8] One gigayear = 1 Gyr = 1 billion years.

and an effective temperature of about 5,300 K for these white dwarfs. Leggett and her collaborators ascribed the chief remaining ambiguities to uncertainties in the core compositions and the unknown masses of the outer layers of white dwarfs and to the effects of phase separation on crystallization of the white dwarf core, which occurs around these faint luminosities.

In 2006 Hugh Harris of the U.S. Naval Observatory and his collaborators determined the WDLF from 6,000 brighter white dwarfs found by the Sloan Digital Sky Survey,[9] obtaining greatly improved statistical accuracy for the higher luminosity end of the luminosity function. They found it to rise smoothly and nearly monotonically to a peak, after which it drops abruptly. As they noted, the precise location of the peak and its shape are actually better determined by the earlier work of Leggett and her coworkers, which focused on the fainter stars. In addition to this peak and subsequent cutoff, Harris and his colleagues noted two other interesting features of their WDLF: (1) There is a slight plateau near $M_{bol} = 10.5$, which may reflect a non-uniform rate of star formation in the galactic disk several gigayears ago; and (2) Just before the peak in the WDLF, there is a slight rise that may reflect the onset of convective coupling between the H-rich envelopes of the white dwarfs and their degenerate cores. Such features begin to hint at other aspects of the white dwarf "fossil record" that may ultimately reveal more than just the age about the early history of the Milky Way Galaxy.

Distances, Ages, and Other Properties of Star Clusters

Stars are formed when dense interstellar clouds of gas and dust contract and fragment under their own self-gravity. We know this because we can see the process going on around us in the Galaxy. For example, the Great Nebula in the constellation Orion—the middle "star" in Orion's sword—is just such a "stellar nursery." It contains a number of very bright, hot stars that brilliantly

[9] Harris, H. C., *et al.* 2006, *Astron. J.*, **131**, 571, "The White Dwarf Luminosity Function from Sloan Digital Sky Survey Imaging Data."

illuminate and ionize the surrounding gas and dust, and observations at infrared wavelengths reveal numerous other stars and protostellar sources that remain shrouded in the cloud. As the stellar radiation ionizes the natal gas cloud, it is gradually cleared away, revealing the cluster of newborn stars.

Star clusters come in all sizes and shapes. The most populous and longest-lived are globular clusters (see Figure 21.3)—roughly spherical agglomerations of hundreds of thousands of stars, which are distributed throughout an extended, spherical halo surrounding the Milky Way Galaxy—and open clusters—irregular groups of some scores to thousands of stars that exist in the galactic disk.

All the stars in a given cluster are formed at approximately the same time, making them excellent "laboratories" for the study of stellar evolution. Graphs of the color magnitude diagrams

FIGURE 21.3 The globular cluster M4. Credit: NOAO/AURA/NSF. Reproduced with permission

(CMDs) for each cluster reveal clear concentrations of stars along the cluster's Main Sequence, and in some cases trace out the red giant branch and the subsequent "blue horizontal branch" as well. Because the cluster stars all lie at roughly the same distance from the Sun, the difference between the apparent and absolute magnitudes is essentially the same for all stars in the cluster, so that a CMD is effectively the same as the Hertzsprung-Russell diagram for that cluster.

A few white dwarfs have long been known to exist in the open cluster called Praesepe and in the Hyades. However, unlike the nearby white dwarfs in the galactic disk, the white dwarfs in star clusters are so distant that it is not possible with current technology to reach all the way to the faintest and oldest of these stars. By the early 1980s, however, CCD detectors had become sufficiently sensitive to enable astronomers to study the brighter white dwarfs as far away as 1,000 pc or more. This made it possible to search for white dwarfs in nearby clusters of stars within the Milky Way Galaxy.

In 1988, Canadian astronomers Harvey Richer and Gregory Fahlman carried out deep CCD photometry of the globular cluster M71,[10] located 3.6 kpc away in the galactic halo, to search for white dwarfs.[11] Confirmed detections of white dwarfs in globular clusters, however, had to await the advent of the *Hubble Space Telescope*, which was capable of reaching much fainter objects than were accessible from the ground.

In 1996, Italian astrophysicist Alvio Renzini and his colleagues obtained *HST* observations of white dwarfs in the globular

[10]The identifying number M71 refers to a specific object listed in a catalog published in 1771 by French astronomer Charles Messier. A dedicated comet hunter, Messier discovered that—unlike comets, which are seen as nebulous patches that move in the sky from night to night—a number of nebulous patches appear in fixed locations in the sky. To avoid "wasting time" on these bothersome and then-unknown objects, he created a list of some 110 of them. The first object in his list, identified as Messier 1 or M1, we now know to be the Crab Nebula, a supernova remnant. Many of the other objects, which were seen as nebulosities because of the relatively poor optical quality of Messier's small telescope, turned out to be star clusters. Others are reflection nebulae—such as the Great Nebula in Orion, M42 and M43—while still others turned out to be planetary nebulae or even external galaxies. The Andromeda Galaxy is M31.

[11]Richer, H. B., and Fahlman, G. G. 1988, *Astrophys. J.*, **325**, 218, "Deep CCD Photometry in Globular Clusters. VI.—White Dwarfs, Cataclysmic Variables, and Binary Stars in M71."

cluster NGC 6752[12] and showed how to use these data to determine the distance to the cluster.[13] They found that the cluster white dwarfs have a very narrow range of masses—(0.53 ± 0.02) M_{Sun}—and follow a very well-defined cooling track in the color magnitude diagram. Because a white dwarf of given mass has essentially a fixed, known radius, the cooling track provides a well-defined relation between white dwarf temperature (or photometric color) and its luminosity or absolute magnitude. By comparing the apparent magnitudes of the cluster white dwarfs with the well-determined absolute magnitudes of nearby white dwarfs in the galactic disk having the same photometric colors, Renzini and his co-workers were accordingly able to determine the distance to this cluster, which they found to be 4.1 kpc. In this way, cluster white dwarfs became useful as "standard candles"—that is, as objects with known intrinsic luminosities—enabling astronomers to determine the distances to star clusters with considerable precision.

Another cluster that has been the focus of appreciable interest is the globular cluster M4. In 1995, Harvey Richer and his collaborators found 258 white dwarfs in this cluster.[14] With masses of (0.51 ± 0.03) M_{Sun}, these white dwarfs also follow a well-defined cooling sequence. In 2002, Brad Hansen and his colleagues studied the white dwarfs in M4 using a very long exposure with the *Hubble Space Telescope*.[15] This enabled them to reach very faint stars, and they were just able to get to the peak of the cluster WDLF and the beginning of its age-determined cutoff. They found the cutoff to be some two magnitudes fainter than the peak in the WDLF for nearby white dwarfs in the galactic disk.

---

[12]More than a century after Messier's catalog appeared, in 1888 the British Royal Astronomical Society published a much more extensive catalog with 7,840 nebulae and star clusters. Called the *New General Catalogue*, it ultimately proved to contain large numbers of external galaxies, and the objects it lists are designated by NGC numbers. Thus, the Crab Nebula is both M1 and NGC 1952, and the Andromeda Galaxy is both M31 and NGC 224. The catalog has been corrected and updated many times, most recently in 1988 on the centenary of its original publication.

[13]Renzini, A., *et al.* **465**, L23, "The White Dwarf Distance to the Globular Cluster NGC 6752 (And Its Age) with the Hubble Space Telescope."

[14]Richer, H. B., *et al.* 1995, *Astrophys. J.*, **451**, L17, "*Hubble Space Telescope* Observations of White Dwarfs in the Globular Cluster M4."

[15]Hansen, B. M. S., *et al.* 2002, *Astrophys. J.*, **574**, L155, "The White Dwarf Cooling Sequence of the Globular Cluster Messier 4."

By fitting both the cluster and disk WDLFs with the same suite of improved white dwarf cooling models, they were thus able to obtain mutually consistent ages of (12.7 ± 0.7) Gyr for M4 and (7.3 ± 1.5) Gyr for the disk. As they pointed out, cluster age determinations from WDLFs are completely independent of those based on the turnoff of stars from the cluster Main Sequences, which have long been used for this purpose. The age Hansen and his co-workers obtained for M4 from the cluster WDLF is consistent with the average age of (13.2 ± 1.5) Gyr obtained from the Main Sequence turnoff for metal-poor globular clusters. It is also consistent with the age of the Universe, which had by then been determined by other methods to be about 14 Gyr, confirming that the oldest globular clusters were among the earliest objects formed after the Big Bang. Four years later, Hansen and his colleagues reported further *HST* observations of white dwarfs in M4, from which they determined the distance to this cluster to be (1.73 ± 0.14) kpc.[16]

In 1996, Adrienne Cool and her colleagues discovered a white dwarf cooling sequence containing about 40 white dwarfs in another globular cluster, NGC 6397.[17] They found these stars to have masses in the range (0.55 ± 0.05) M_{Sun}. A decade later, Hansen and his colleagues also investigated this cluster using their improved white dwarf models.[18] Using Renzini's white dwarf fitting technique, they found the distance to the cluster to be 2.5 kpc and determined the cluster age to be 11.47 ± 0.47 Gyr. This is in good agreement with the age obtained from the cluster turnoff luminosity, where stars leave the Main Sequence on their way to becoming red giants.

Such distance and age determinations obtained from the cluster white dwarf stars support an emerging picture of the formation and evolution of the Milky Way Galaxy. In this cosmological scenario, the Big Bang occurred about 14 billion years ago, forming the Universe in which we live. The Universe has been expanding and cooling ever since. The inexorable pull of gravity caused slightly denser concentrations of mass to begin to contract. Perhaps

---

[16] Hansen, B. M. S., *et al.* 2006, *Astrophys. J. Suppl.*, **155**, 551, "Hubble Space Telescope Observations of the White Dwarf Cooling Sequence of M4."

[17] Cool, A. M., Piotto, G., and King, I. R. 1996, *Astrophys. J.*, **468**, 655, "The Main Sequence and a White Dwarf Sequence in the Globular Cluster NGC 6397."

[18] Hansen, B. M. S., *et al.* 2007, *Astrophys. J.*, **671**, 380, "The White Dwarf Cooling Sequence of NGC 6397."

a billion years after the Big Bang, this led to the formation of the halo and globular clusters in our Galaxy.

Under the combined influence of gravity and rotation, infalling gas continued to settle into a hot, extended galactic "protodisk." The gaseous protodisk continued to cool and settle toward the galactic plane for another 4–5 billion years, at which time vigorous star formation began in the thin disk, producing additional stars that ultimately evolved into the population of nearby white dwarfs we see around us in the disk today.

The WDLFs of globular clusters also are beginning to be useful in constraining other physical properties of astronomical objects. In 2009, Don Winget and his colleagues pointed out the existence of a peak in the cluster WDLF of NGC 6397 that occurs at luminosities above the cutoff due to the cluster age.[19] Analyzing white dwarfs near this peak, they showed that the data are consistent with white-dwarf crystallization theory and confirm that the liquid-solid transition in a white dwarf core is a first-order phase transition, releasing latent heat that detectably prolongs the cooling at low luminosities. Winget and his colleagues also showed how improved observational data can be used in the future to constrain the temperature at which crystallization occurs as well as the C/O ratio in white dwarf cores.

[19] Winget, D. E., *et al.* 2009, *Astrophys. J.*, **693**, L6, "The Physics of Crystallization from Globular Cluster White Dwarf Stars in NGC 6397."

22. White Dwarfs and Cosmology

In 2011, Saul Perlmutter, Adam Riess, and Brian P. Schmidt (Figure 22.1) won the Nobel Prize for their work in demonstrating that the expansion of the Universe is accelerating. This was a complete surprise, as astronomers had generally expected the expansion either to continue at a steady pace or to slow down gradually. How did Perlmutter and his colleagues make this amazing discovery, and what does it imply for our understanding of the nature of the Universe? And what can white dwarf stars *possibly* have to do with this?

To answer the first question, it may be helpful to put this work in context. Almost a century before the astonishing discovery by Perlmutter and his colleagues, American astronomer Henrietta Leavitt—working at the Harvard College Observatory in 1912—set the stage for subsequent advances in cosmology, that branch of the physical sciences that concerns the origin and evolution of the Universe. At the time, she was studying a particular class of pulsating stars called Cepheid variables. These stars undergo regular oscillations, brightening and dimming with well-defined and easily measured pulsation periods.

From her careful work, Leavitt was able to show that there is a direct relationship between the period of the oscillation and the intrinsic luminosity of the star. The longer the period, the brighter the star.[1] This immediately made it possible for astronomers to use Cepheid variables as "standard candles" to determine accurate distances to remote astronomical objects. The particular advantage of the Cepheids is that they are very bright stars—thousands to tens of thousands of times more luminous than the Sun—and thus that they can be seen out to relatively large distances.

[1] Leavitt, H. S., and Pickering, E. C. 1912, *Harvard Coll. Obs. Circ.*, **173**, 1, "Periods of 25 Variable Stars in the Small Magellanic Cloud."

H.M. Van Horn, *Unlocking the Secrets of White Dwarf Stars*, Astronomers' Universe, DOI 10.1007/978-3-319-09369-7_22

FIGURE 22.1 Saul Perlmutter, Adam Riess, and Brian Schmidt during a press conference on the occasion of winning the Nobel Prize in 2011. Public domain image courtesy Wikimedia Commons (http://upload.wikimedia.org/wikipedia/commons/9/9e/Nobel_Prize_2011-Press_Conference_KVA-DSC_7900.jpg)

A few years later, Albert Einstein published his general relativistic theory of gravitation,[2] for the first time providing a framework capable of supporting a scientific inquiry into the nature of the Universe. At the time, astronomers believed that the Universe was static and unchanging, so Einstein added a term to his equations that he called the "cosmological constant," which did allow static solutions.

During the 1920s, American astronomer Edwin P. Hubble, then working at the 100-in. (2.5-m) telescope on Mt. Wilson, used the Cepheids to show that in fact the galaxies in the Universe are all rushing away from each other at great speeds and that the more distant they are, the faster they are moving.[3] After this discovery, Einstein came to regard the cosmological constant as "the

[2] Einstein, A. 1915, *Sitzungsberichte de Königlich Preußischen Akademie der Wissenschaften (Berlin)*, Seite 844—847, "Die Feldgleichungen der Gravitation."

[3] Hubble, E. 1929, *Publ. Nat. Acad. Sci.*, **15**, 168, "A Relation Between Distance and Radial Velocity among Extragalactic Nebulae."

biggest blunder of my life."[4] The constant of proportionality in the linear relationship between distance and recession velocity that Hubble discovered—now called the "Hubble constant," H_0—unfortunately is very difficult to determine accurately. American astronomer Alan Sandage, himself one of the preeminent leaders in this effort, subsequently described observational cosmology by the latter part of the twentieth century as a quest for two numbers—the Hubble constant and a dimensionless "deceleration parameter"—that were frustratingly just beyond reach.

Another step in scientists' growing understanding of the Universe came the year after the end of World War II, when physicist George Gamow showed that the observed cosmic expansion implies that the Universe must originally have existed in a state with vastly greater temperatures and densities than exist today and that it must then have been dominated by thermal radiation.[5]

Because the wavelengths of thermal radiation lengthen as the Universe expands, the temperature of the cosmic background radiation would be only a few degrees Kelvin today, the peak of the spectrum lying in the range of centimeter-wavelength microwaves. In 1965, physicists Arno Penzias and Robert W. Wilson—then working at Bell Telephone Laboratories in Holmdel, New Jersey, on a project to reduce or eliminate sources of radio noise in microwave communication—first detected this cosmic microwave background (CMB) radiation at wavelengths corresponding to a temperature of about 3 K.[6] In 1978, they were awarded the Nobel Prize in Physics for this work. Their discovery confirmed what has come to be called the "Big Bang" model for the expanding Universe. That is, that the Universe began at unimaginably high densities and temperatures at a singular moment in the distant past and has been expanding ever since.

On the theoretical side, cosmological models based on general relativity and including only radiation and ordinary matter

[4] Misner et al. (1973), p. 707.

[5] Gamow, G. 1946, *Phys. Rev.*, **70**, 572, "Expanding Universe and the Origin of Elements."

[6] Penzias, A. A., and Wilson, R. W. 1965, *Astrophys. J.*, **142**, 419, "A Measurement of Excess Antenna Temperature at 4,080 Mc/s."

(termed "baryons") ran into difficulties in trying to explain the formation of galaxies. This problem could be alleviated, however, if most of the matter in the Universe, instead of being ordinary matter, as had up to then been assumed, were instead the mysterious "dark matter" that had been inferred from astronomical measurements as early as the 1930s.

The first truly convincing evidence for this still-mysterious substance came only in the 1970s, when American astronomer Vera Rubin and her colleague Kent Ford, working at the Department of Terrestrial Magnetism of the Carnegie Institution of Washington, in Washington, D.C., showed that the orbital motions of stars around the centers of spiral galaxies required the presence of much more mass than could be provided by visible matter (stars and gas and dust) in the galaxies.[7] Additional evidence for the existence of dark matter soon followed, and today, the evidence for the existence of dark matter in galaxies or larger cosmic structures has become overwhelming, although we still do not know what it is.

During the 1980s, large-scale surveys of Galaxy redshifts—such as that by Harvard's Margaret Geller and the late John Huchra (1948–2010)—began to reveal morphological details of the cosmological distribution of galaxies, showing that on very large scales the distribution of matter has a filamentary or cellular structure.[8] In the current standard model of cosmology, such structures form along filaments of dark matter, which are traced out by the galaxies and clusters of galaxies they contain.

A further advance in the observational basis for cosmology was provided in 1989, when the *Cosmic Background Explorer (COBE)* spacecraft was launched with a 4-year mission to map the cosmic microwave background with exquisite precision. Astronomers working with *COBE* soon showed that the observed spectrum of the CMB fits a black-body thermal radiation spectrum with a temperature of 2.73 K to extraordinarily high preci-

---

[7] Rubin, V. C., and Ford, W. K. 1970, *Astrophys. J.*, **159**, 379, "Rotation of the Andromeda Nebula from a Spectroscopic Survey of Emission Lines." See also Rubin, V. C., Ford, W. K., Jr., and Thonnard, N. 19870, *Astrophys. J.*, **238**, 471, "Rotational Properties of 21 Sc Galaxies with a Large Range of Luminosities and Radii, from NGC 6405 (R = 4 kpc) to UGC 2885 (R = 122 kpc)."

[8] Geller, M. J., and Huchra, J. P. 1989, *Science*, **246**, 897, "Mapping the Universe."

sion.[9] There are minute fluctuations in the CMB, but only at the level of about one part in 100,000. Nevertheless, the structures we see around us in the Universe today have grown from these tiny "seeds." Astronomers George Smoot and John Mather, who were principle investigators for the *COBE* mission, were awarded the Nobel Prize in 2006 for this important work.

Let us now return to the scientific advance made by Perlmutter and his colleagues. What exactly did they do, and how did their work affect our understanding of the Universe?

To measure the distances to galaxies at truly cosmological locations, where the redshifts[10] z produced by the expansion of the Universe approach and even exceed unity, astronomers had to find a new type of "standard candle" that could be seen to much larger distances than the Cepheid variables. Even the brightest Cepheids can be detected only out to a few tens of megaparsecs. Fortunately, another class of objects did gradually come to be useful for this purpose: the exploding stars called supernovae. These are stellar explosions that release so much energy that for about a month they rival the luminosity (power output) of all the hundreds of billions of stars in an entire Galaxy! The energy release is so enormous that it generally disrupts the star completely.

Observational studies—such as those carried out by Fritz Zwicky[11] and his colleagues at Caltech around the middle of the twentieth century—revealed several different types of supernovae (SNe) in distant galaxies. Each supernova was characterized by the shape of its light curve, as its luminosity declines following the outburst; by the nature of the host Galaxy in which the supernova occurs (*e.g.*, spiral or elliptical); and by properties such as the chemical composition of the Galaxy. Supernovae also differ in the way their spectra appear and change over time as the outburst gradually fades. Type II supernovae display hydrogen in their spectra, for example, while Type I supernovae—subdivided into Types Ia, Ib, and Ic—do not. The light curves of SNe Ia are all remarkably

---

[9] Bennett, C. L., *et al.* 1996, *Astrophys. J.*, **464**, L1, "Four-Year COBE DMR Cosmic Microwave Background Observations: Maps and Basic Results."

[10] The redshift z is defined as the amount of change in the wavelength of a spectral line produced by the cosmological recession speed divided by the un-shifted wavelength of the line.

[11] Zwicky, F. 1940, *Revs. Mod. Phys.*, **12**, 66, "Types of Novae."

similar, as shown by University of Oklahoma astrophysicist David Branch and his colleagues in the early 1990s,[12] and they also are intrinsically the brightest of the supernovae.

Before supernovae could be employed as standard candles for cosmology, however, it was first necessary for the Supernova Cosmology Project led by Saul Perlmutter at Lawrence Berkeley Laboratory and the competing High-*z* Supernova Search Team led by Robert P. Kirshner at Harvard—of which Adam Riess and Brian Schmidt were members—to calibrate the luminosities of these stellar explosions. They did this in several different ways.[13] For supernovae in relatively nearby galaxies, they could determine the distances using the period-luminosity relation for the very bright Cepheid variable stars in the same Galaxy. They also employed a correlation that had been established between the peak luminosity of a SN Ia and the shape of its light curve. A third method involved the correlation between features seen in the spectrum of a Type Ia supernova and its maximum brightness. The combination of such different methods provided confidence in the use of SNe Ia as calibrated "standard candles" for cosmological investigations, a possibility that had been foreseen decades earlier but which had to await the advance of technology before these exploding stars could be so employed.

What could the very faint white dwarf stars possibly have to do with supernovae that are bright enough to be seen halfway across the Universe? The answer is that these faint stars are thought to be the progenitors of SNe Ia. As we have seen, white dwarfs are extremely degenerate objects that consist mainly of carbon and oxygen. At temperatures of a few tens of millions of degrees Kelvin, the cores of these stars are far too cool to support thermonuclear reactions among the nuclei. If the white dwarf is a member of a close binary system, however, it may accrete matter transferred from its companion and increase in mass.

As the white dwarf gains mass and approaches the Chandrasekhar limit—about $1.4\,M_{\mathrm{Sun}}$—its radius decreases and its central density increases. This forces the C and O nuclei in the center

---

[12] Branch, D., and Miller, D. L. 1993, *Astrophys. J.*, **405**, L5, "Type Ia Supernovae as Standard Candles."

[13] Perlmutter, S., *et al.* 1997, *Astrophys. J.*, **483**, 565, "Measurements of the Cosmological Parameters Ω and Λ from the First Seven Supernovae at $z \geq 0.35$."

of the star to move closer and closer together. Eventually, the nuclei come so close together that so-called "pycnonuclear" reactions[14] can occur, even at the very low temperatures of the white dwarf cores. Reactions such as $^{12}C + ^{12}C \rightarrow ^{24}Mg$, $^{12}C + ^{16}O \rightarrow ^{28}Si$, and $^{16}O + ^{16}O \rightarrow ^{32}S$ then occur, releasing energy in the process.

The onset of these reactions marks the beginning of the end for the white dwarf. In a non-degenerate star such as the Sun, if the rate of nuclear energy production in the core were to increase slightly for some reason, thus raising the core temperature, the resulting increase in pressure would cause the star to expand slightly, reducing the temperature and density and causing the energy-generation rate to decline again to its original level. In other words, the effect is like that of a thermostat, with slight expansions or contractions of the star able to keep the temperature, density, and nuclear energy production rate in stable equilibrium. In a degenerate star such as a white dwarf, however, this mechanism no longer operates. A slight increase in nuclear energy production does indeed produce a corresponding slight increase in the temperature. But in strongly degenerate matter, the pressure does not undergo a corresponding increase, so the star does not expand and reduce its rate of nuclear energy generation. Instead, the increase in temperature causes a further increase in nuclear energy production, leading to a full-scale thermonuclear runaway.

By the time the thermal energy has increased enough to enable the star finally to expand—becoming comparable to the Fermi energy of the degenerate electrons, at a temperature of perhaps 10 billion degrees Kelvin—it is high enough so that thermonuclear reactions have already consumed all of the available nuclear fuel, leaving behind newly formed chemical elements near the so-called "iron peak" in the Periodic Table,[15] primarily radioactive ^{56}Ni.

[14] "Pycnonuclear" reactions are density-induced nuclear reactions, as opposed to temperature-induced "thermonuclear" reactions.

[15] The iron found on Earth today thus originated inside an ancient C/O white dwarf, which exploded as a Type Ia supernova billions of years ago, dispersing out into space matter that was subsequently incorporated into the gas and dust from which our Solar System formed about 5 billion years ago. Indeed, astronomers think that SNe Ia create much of the iron in the Universe. Elements with atomic masses greater than those of the iron-peak elements are not created in energy-*producing* nuclear reactions, because iron is the most strongly bound atomic nucleus. To produce still heavier elements requires putting energy *into* the reactions through a number of different processes, rather than getting energy out.

In the process, the white dwarf is completely incinerated, releasing so much energy that the star is explosively disrupted. The initial rise to a peak brightness about five billion times as luminous as our Sun takes about 3 weeks. The maximum luminosity is about the same for all Type Ia supernovae, presumably because the masses of the exploding white dwarfs are all about the same—the Chandrasekhar limit—and this is the feature that makes SNe Ia useful as standard candles for cosmology.

And what did Perlmutter and his colleagues learn about the nature of the Universe from these exploding stars? By the end of the twentieth century, after a decade of work, the Supernova Cosmology Project had found 75 Type Ia supernovae with redshifts in the range $z \approx 0.2$–0.9. All of the supernovae had been found, before reaching maximum light, using the 4-m telescope at the Cerro Tololo Interamerican Observatory in Chile. The supernovae were subsequently followed photometrically for the next 40–60 days using various 4-m class telescopes, while the redshifts and spectral types were obtained using the Keck I and II 10-m telescopes atop Mauna Kea in Hawaii.[16]

The Hubble diagram for these supernovae—a graph of apparent magnitude as a function of the redshift z—clearly demonstrates that supernovae at higher redshifts are systematically fainter than expected. Since the redshift directly gives the recession velocity, while the "standard candles" provided by the luminosities of the SNe Ia independently measure the distance, this implies that the supernovae with higher redshifts are farther away than expected and thus that the expansion of the Universe is accelerating.

This startling result led Perlmutter and his colleagues to conclude that only about 28 % of the total mass-energy density of the Universe is composed of ordinary matter plus dark matter.[17] That is, most of the mass-energy of the Universe appears to consist of some even more mysterious "dark energy," generally represented by a "cosmological constant," denoted by the Greek letter lambda (Λ), in Einstein's general relativistic theory of gravitation. It is

[16] Perlmutter, S., *et al.* 1999, *Astrophys. J.*, **517**, 565, "Measurements of Ω and Λ from 42 High-Redshift Supernovae."
[17] *Ibid.*

important to note that this dark energy is completely distinct from—and should not be confused with—dark matter. Perlmutter and his team checked carefully to eliminate or account for errors, and they were able to show that their conclusions are "robust" (scientists' jargon, meaning that they have great confidence in their results). The investigators also found the expansion age of the Universe to be 14.9 Gyr.

At about the same time, Riess, Schmidt, and their colleagues on the competing High-z Supernova Search Team reported their own results.[18] They studied ten SNe Ia found in the range of redshifts $z \approx 0.2–0.6$, using similar methods to those of the Supernovae Cosmology Project. They also found the distances to high-redshift supernovae to be some 10–15 % larger than expected. Using various methods of analysis, they obtained results similar to those of Perlmutter's group. Having two independent—and indeed, competing—teams arrive at essentially identical results provides confidence that the stunning conclusions are correct.

After the discovery of the accelerating expansion of the Universe from measurements of Type Ia supernovae, a cosmological model called the "Lambda Cold Dark Matter," or ΛCDM, quickly became the standard.[19] This is a parametrized Big Bang cosmological model founded on Einstein's general relativistic theory of gravitation that includes both cold dark matter and a cosmological constant Λ representing the dark energy hypothesized to account for the accelerating expansion of the Universe. It is the simplest model that fits the existing observations of the cosmic microwave background, including the very small-amplitude variations in the temperature; the large-scale structure in the cosmological distribution of galaxies; the abundances of hydrogen (^{1}H), deuterium (^{2}H), helium-three (^{3}He), helium (^{4}He), and lithium (^{7}Li)—all of which were produced by nuclear reactions during the first few minutes after the Big Bang—and the accelerating expansion of the Universe as determined from distant supernovae.

[18] Riess, A. G., *et al.*, *Astron. J.*, **116**, 1009, "Observational Evidence from Supernovae for an Accelerating Universe and a Cosmological Constant."

[19] See, for example, Spergel, D. N., *et al.*, *Astrophys. J. Suppl.*, **170**, 377, "Three-Year *Wilkinson Microwave Anisotropy Probe (WMAP)* Observations: Implications for Cosmology."

Fitting the ΛCDM cosmological model to the observational data available in 2007 yields the age of the Universe as 13.7 Gyr, and the Hubble constant as $H_0 = 73$ km s^{-1} Mpc^{-1}. In addition, investigators found that the fraction of the mass-energy density of the Universe due to baryons is only about 4 %, while that due to cold dark matter is about 20 %, and the contribution due to dark energy—as represented by the cosmological constant Λ—is a whopping 73%! The model also gives the age of the Universe at the time radiation decoupled from matter to be about 300,000 years after the Big Bang. The redshift at this early epoch is $z \approx 1{,}100$, while the redshift at the time when the first stars and accreting supermassive black holes in active galactic nuclei re-ionized the Universe is $z \approx 10$.

These results are by no means the last word in observational cosmology, as the rate at which new data and new theories are becoming available continues to increase. Indeed, cosmology has become a very active area of research, and the goal of improving our understanding of the very early stages in the evolution of the Universe remains a major challenge to astronomy and physics today. However, this is beyond the scope of the present book. And we still do not know what constitutes the dark matter that appears to be present in large amounts throughout the Universe, and dark energy is even more mysterious.

We have now reached the end of our story about white dwarf stars. This is not the final chapter in the saga, however. Indeed, the study of a few curiously anomalous, faint stars known a century ago has today matured into a full-fledged subdiscipline of astronomy and astrophysics. The remarkable way in which continuing advances in observational astronomy, theoretical astrophysics, physics, and instrumentation have combined to make possible our present understanding of white dwarf stars provides a superb case study of the way in which science advances through such interconnected developments. And as we have seen, our increased understanding of white dwarfs has made it possible to utilize them to make new types of astronomical measurements.

Fresh observational discoveries, continuing advances in theory and computation, and progress in the development of technology and observing techniques continues to expand our understanding of, and knowledge about, the white dwarf stars. And as longstanding puzzles are solved, new mysteries continue to pop up, providing further challenges for the years ahead.

The most recent Workshop at the time of this writing was held at the Universite de Montreal in 2014, the 40th anniversary of the initial meeting, and attracted more than 130 participants from more than 20 nations. While many of the general topics on the agenda of the Montreal Workshop were the same as those at the very first in the series, the contents of the presentations represented the most up-to-date research. And new topics continue to be added, which could not even have been imagined four decades ago.

A measure of the current vitality of white dwarf research is provided by the European Workshops on White Dwarfs. Initiated at the Universitaet Kiel by Volker Weidemann in 1974 with perhaps a dozen participants, these biennial meetings have since grown into the major international conferences on white dwarfs and have been hosted at venues around the globe.

Those of us who have been privileged to take part in this fascinating journey of discovery have enjoyed both the camaraderie and the friendly competition with our colleagues around the world as we have worked together to unlock many of the secrets of the white dwarf stars. We are impressed and inspired by the innovative approaches being taken by the current generation of white-dwarf researchers in tackling the challenges being presented by these fascinating objects.

Appendix A: Some Useful Astronomical Units

A.1 Distances

Because astronomical distances are so much larger than any distance we encounter on Earth, it is useful to have some special units to represent them. For example, the distance from Earth to the Sun provides a convenient unit of distance within the Solar System. Called the astronomical unit, or AU, that distance is

$$1\,\mathrm{AU} = 1.496 \times 10^{8}\,\mathrm{km} = 92.96 \times 10^{6}\,\mathrm{miles}.$$

To sufficient accuracy for our present purposes, we can write this as

$$1\,\mathrm{AU} = 150\,\mathrm{million\,km} = 93\,\mathrm{million\,miles}.$$

Even this large a distance is not big enough to serve as a convenient measure of the distances between the stars. A quantity that *is* large enough is the distance a beam of light travels in one year—a distance known as a light year. Since the speed of light has been measured to be about 186,000 miles per second—or equivalently about 300,000 km s^{-1}—and since there are about 31,560,000 s in a year, a light year is a distance of 300,000 km s^{-1} times 31,560,000 s, which equals 9.47×10^{12} km or about 5.87×10^{12} miles. This is about 6.33×10^{4} AU. That is, a star located one light year from the Sun is about 63,000 times farther away than is Earth.

For practical observational reasons, astronomers tend to prefer another unit of distance called the parsec, abbreviated pc. This turns out to be about 3.26 light years, or 3.09×10^{13} km. Some objects within our own Milky Way Galaxy lie thousands of parsecs away from the Solar System, making a more convenient distance unit for such objects the kiloparsec, abbreviated kpc: 1,000 pc = 1 kpc. For example, the center of our Galaxy lies about 8.5 kpc

H.M. Van Horn, *Unlocking the Secrets of White Dwarf Stars*, Astronomers' Universe, DOI 10.1007/978-3-319-09369-7

away from the Sun, and the radius of the galactic disk is about 20–25 kpc. Other galaxies beyond the Milky Way lie at even greater distances. The Andromeda Galaxy, which may be similar in size and shape to the Milky Way, lies at a distance of about 725 kpc. This makes a million parsecs—called a megaparsec and abbreviated Mpc—a more practical unit of distance measurement in the extragalactic realm.

And as a result of astronomer Edwin Hubble's discovery early in the twentieth century that more distant galaxies are speeding away from us at ever-faster speeds, the redshift in the light from these sources has become the standard for distance measurement in the remotest parts of the Universe. But that is another story.

The different units for astronomical distance measurements are summarized in the following table:

1 AU = 1.5×10^8 km
1 light year = 9.5×10^{12} km = 6.3×10^4 AU
1 pc = 3.3 light years = 3.1×10^{13} km
1 kpc = 1,000 pc = 3.1×10^{16} km
1 Mpc = 1,000,000 pc = 3.1×10^{19} km

A.2 Stellar Parameters

Just as special units are helpful in dealing with astronomical distances, so, too are different units useful in discussing stars. In this case, our own star—the Sun—turns out to be a pretty ordinary star, and it is useful to express the properties of other stars in terms of solar values. The quantities most often used are the mass of the Sun, represented by the symbol M_{Sun}; its radius, R_{Sun}; and the total solar power output, called the luminosity, L_{Sun}. In terms of conventional physical units, these quantities are:

$M_{Sun} = 1.99 \times 10^{30}$ kg
$R_{Sun} = 6.96 \times 10^5$ km
$L_{Sun} = 3.90 \times 10^{26}$ W

For purposes of comparison, the mass of the Sun is more than 300,000 times that of Earth and about 1,000 times larger than the mass of Jupiter, the largest planet in the Solar System. Similarly, the radius of the Sun is about 100 times larger than Earth's and about ten times larger than Jupiter's. And the total power output of the Sun is about 400 billion billion times larger than a 1 MW generating station.

It may also be helpful to get a sense of the way the Sun compares with other stars. The smallest Main Sequence stars (hydrogen-burning stars such as the Sun)—called "red dwarfs"—have masses and radii of about 0.1 M_{Sun} and 0.1 R_{Sun}, respectively, while their luminosities are only about 0.001 $L_{\mathrm{Sun}} = 10^{-3} L_{\mathrm{Sun}}$. Conversely, the most massive stars may approach or even exceed 100 M_{Sun}. The stars with the largest dimensions are red giants and supergiants, which may have radii as large as 1,000 $R_{\mathrm{Sun}} = 10^3 R_{\mathrm{Sun}}$. And the very brightest stars may have luminosities of 100,000 $L_{\mathrm{Sun}} = 10^5 L_{\mathrm{Sun}}$ or more. As a specific example, the bright, blue-white star Sirius has a mass of 2.0 M_{Sun}, radius 1.7 R_{Sun} and luminosity of 25 L_{Sun}. As another example, the red supergiant Betelgeuse—the brightest star in the constellation Orion—has a mass about 20 M_{Sun}, a radius about 800 R_{Sun}, and a luminosity of about 50,000 L_{Sun}.

Appendix B: Powers of Ten and Logarithms

Astronomical dimensions are much larger than those we encounter in our everyday lives here on Earth. Consequently, when they are expressed in familiar units, they are represented by very large numbers. For example, the distance from Earth to the Sun is approximately 93 million miles, or about 150 million km. If all the decimal places are written out, the latter figure becomes 150,000,000 km, or $1.5 \times 100{,}000{,}000$ miles. Instead of writing out all these zeros for each large number, scientists long ago adopted the practice of writing it in terms of a number of powers of ten: $100{,}000{,}000 = 10 \times 10 \times 10 \times 10 \times 10 \times 10 \times 10 \times 10 = 10^8$. That is, you need to multiply eight factors of ten together to get the number 100 million. This is clearly a much more compact way to write a very large number.

It is often useful as well to be able to talk about just the power of ten itself, rather than the entire number. This quantity is called the logarithm of the number. In our example using the number 100 million, the logarithm of 100 million equals logarithm $(10^8) = 8$. For simplicity, logarithm is generally abbreviated to just log, giving log $(10^8) = 8$.

Many hand-held electronic calculators will return the logarithm of a quantity at the touch of a button—or perform the reverse operation of returning the number, given the logarithm. If this is sufficient for your interest, the remainder of this appendix can be safely ignored. Conversely, if you *are* interested in learning more, read on.

One of the advantages of using logarithms is that multiplication becomes just the addition of logarithms. For example, consider a hypothetical object at a distance of 100×100 million km. In powers of ten, this distance would be written as 100×100 million $= 10^2 \times 10^8 = 10^{10}$. The logarithm of this quantity is log $(100 \times 100 \text{ million}) = \log(10^2 \times 10^8) = \log(10^2) + \log(10^8) = 2 + 8 = 10$.

H.M. Van Horn, *Unlocking the Secrets of White Dwarf Stars*, Astronomers' Universe, DOI 10.1007/978-3-319-09369-7

So, multiplication of numbers is equivalent to addition (or subtraction) of logarithms.

Just as positive logarithms represent multiplication by powers of ten, so negative logarithms represent division by powers of ten. Again using our example, a distance equal to one-tenth of 100 million km becomes $(1/10)\times10^8=10^7$. Or: $\log(10^7)=\log([1/10]\times10^8)=\log(1/10)+\log(10^8)$. Thus, $\log(1/10)=\log(10^7)-\log(10^8)=7-8=-1$.

This also provides an easy way for us to see that the logarithm of the number one is zero: $\log(1)=\log(([1/10]\times10)=\log(1/10)+\log(10)=-1+1=0$.

Of course, there are many natural numbers between one and ten—for example, 2 or 6.5 or 3.14159 ... Can they be represented by logarithms, too? The answer is "Yes."

One way to see this is to consider square roots, cube roots, *etc.* Suppose that "x" represents the square root of 10. That is, $x\times x=10$. (The actual numerical value of the square root of 10 is 3.1622 ... , but that doesn't matter for our present purposes.) By taking the logarithm of both sides of this equation, we get $\log(10)=\log(x^2)=\log(x)+\log(x)=2\log(x)$. But since $\log(10)=1$, this means that $\log(x)=½=0.5$. Our original equation gave $x=\sqrt{10}=10^{0.5}$, and this is the same result we get from the logarithm: $x=10^{\log(x)}=10^{0.5}$. Making use of the actual numerical value of the square root of 10, we have $3.1622\ldots=10^{0.5}$, or equivalently, $0.5=\log(3.1622\ldots)$.

As another example, suppose that "y" is the cube root of 10. Then $y\times y\times y=y^3=10$. This gives $y=10^{1/3}=10^{0.333\ldots}$. And using the numerical value of the cube root of 10, we have $y=10^{0.333\ldots}=2.1544\ldots$ In terms of logarithms, this gives $\log(2.1544\ldots)=0.333\ldots$

In fact, the logarithm of any number between 1 and 10 turns out to be a number between 0 and 1, and this gives us a very simple way to represent very large numbers. To go back to our original example, $\log(150{,}000{,}000)=\log(1.5\times10^8)=\log(1.5)+\log(10^8)=0.1760\ldots+8=8.1760\ldots$

Appendix C: Chandrasekhar's Models for Fully Degenerate White Dwarfs

The following table summarizes the properties of Chandrasekhar's fully degenerate (*i.e.*, zero-temperature) models for white dwarf stars, as computed from his 1939 book. The quantity ρ_c is the central mass density in grams per cubic cm, while P_c is the central pressure in dynes per square cm.[1] The quantities M/M_{Sun} and R/R_{Sun} are the mass and radius of the model in units of the corresponding solar quantities. The models have masses roughly comparable to that of the Sun, but the radii are about 100 times smaller—or even less—and are more nearly comparable to the radius of Earth. The last two columns give the mean density of the model, again in grams per cubic cm, and the ratio of the central density to the mean density.

ρ_c (g cm^{-3})	P_c (dyn cm^{-2})	M/M_{Sun}	R/R_{Sun}	$\langle\rho\rangle$ (g cm^{-3})	$\rho_c/\langle\rho\rangle$
2.46×10^{5}	2.77×10^{21}	0.220	2.00×10^{-2}	3.84×10^{4}	6.40
1.07×10^{6}	2.88×10^{22}	0.405	1.54×10^{-2}	1.54×10^{5}	6.94
1.96×10^{6}	7.38×10^{22}	0.505	1.38×10^{-2}	2.68×10^{5}	7.43
3.60×10^{6}	1.86×10^{23}	0.612	1.23×10^{-2}	4.58×10^{5}	7.85
7.00×10^{6}	4.97×10^{23}	0.737	1.08×10^{-2}	8.08×10^{5}	8.67
1.57×10^{7}	1.60×10^{24}	0.885	9.25×10^{-3}	1.58×10^{6}	9.94
5.30×10^{7}	8.87×10^{24}	1.081	7.11×10^{-3}	4.20×10^{6}	12.6
1.63×10^{8}	4.13×10^{25}	1.219	5.51×10^{-3}	1.02×10^{7}	16.0
6.74×10^{8}	2.83×10^{26}	1.330	3.90×10^{-3}	3.14×10^{7}	21.4
1.93×10^{9}	1.17×10^{27}	1.375	2.96×10^{-3}	7.40×10^{7}	26.1

In the models with lower masses, the degenerate electrons are non-relativistic. That is, the speeds of the electrons at the Fermi energy are less than the speed of light. The electrons begin to reach

[1] A pressure of 6.9×10^{4} dyn cm^{-2} equals 1 pound per square inch, or 1 PSI.

H.M. Van Horn, *Unlocking the Secrets of White Dwarf Stars*, Astronomers' Universe, DOI 10.1007/978-3-319-09369-7

relativistic speeds in models with masses around $0.8\,M_{Sun}$, and they become increasingly relativistic as the central densities increase. The radii of the models become smaller and smaller as the masses approach the Chandrasekhar limit, about $1.4\,M_{Sun}$.

For comparison, the density of lead is about 11 g cm^{-3}, while that of uranium is about 20 g cm^{-3}. A more appropriate comparison is with the density and pressure at the center of the Sun, which are about 160 g cm^{-3} and 2.2×10^{17} dyn cm^{-2}, respectively. The central density and pressure of even the least-massive model listed in the table, with $0.220\,M_{Sun}$, are respectively more than 1,000 and 10,000 times greater than these values. And for a 1.0 M_{Sun} white dwarf such as Sirius B, the central density reaches almost 40 million grams per cubic cm, about 250,000 times greater than that of the Sun, and the central pressure is some 6×10^{24} dyn cm^{-2}, more than 30 million times the solar value.

Appendix D: "WD Numbers" for White Dwarfs in this Book

α Canis Majoris B	WD0642-166	(Also called Sirius B, EG 049, LFT 0486)
α Canis Minoris B	WD0736+053	(Also called Procyon B)
40 Eridani B	WD0413-077	
BPM 25114	WD1743-521	
EC 20058–5234	WD2006-523	
EUVE J0317–853	WD0317-855 J	(Also called REJ0317-853, LB 9802)
Feige 4	WD0017+136	
Feige 7	WD0041-102	
Feige 24	WD0232+035	
G191–B2B	WD0501+527	
GD 52	WD0348+339	
GD 154	WD1307+354	
GD 358	WD1645+325	(Also called V477 Herculis)
Grw +70° 8247	WD1900+705	
H1504+65	WD1501+664	
HD149499B	WD1634-573	
HL Taurus 76	WD0416+272	
HZ 43	WD1314+293	
L 19–2	WD1425-811	
L 879–14	WD0435-088	
LB 1497	WD0349+247	
MCT 0455–2812	WD0455-282	
PG 1031+234	WD1031+234	
PG 1159–035	WD1159-034	(Also called GW Virginis)
PG 2131+066	WD2131+066	
R 548	WD0133-116	(Also called ZZ Ceti)
Stein 2051 B	WD0431+58x	(Also called G175-034, LHS 26/27)

The WD numbers listed in this table are mainly from McCook, G. P., and Sion, E. M. 1999, *Astrophys. J. Suppl.*, 121, 1, "A Catalog of Spectroscopically Identified White Dwarfs."

H.M. Van Horn, *Unlocking the Secrets of White Dwarf Stars*, Astronomers' Universe, DOI 10.1007/978-3-319-09369-7

Abbreviations and Symbols

ABC	Atanasoff-Berry computer
AGB	Asymptotic giant branch
AM CVn	AM Canum Venticorum variable
ANS	*Astronomical Netherlands Satellite*
AU	Astronomical unit
BPM	Bruce Proper-Motion survey
CCD	Charge-Coupled Device
CDC	Control Data Corporation
CGS	Centimeter-gram-second
CMB	Cosmic microwave background
CMD	Color magnitude diagram
CNO	Carbon, nitrogen, oxygen
CNRS	Centre National de la Recherche Scientifique
C/O or CO	Carbon-oxygen mixture
COBE	*Cosmic Background Explorer*
CoRoT	*Convection, Rotation, and planetary Transients spacecraft*
COSTAR	*Corrective Optics Space Telescope Axial Replacement*
CRAQ	Centre de Recherche en Astrophysique du Québec
CSPN	Central star of planetary nebula
CTIO	Cerro Tololo Interamerican Observatory
CTR	Computing, Tabulating, Recording Corporation
DAV	Variable DA white dwarf
DBV	Variable DB white dwarf
DOV	Variable DO white dwarf
EC	Edinburgh-Cape survey
EDVAC	Electronic Discrete Variable Automatic Computer
ENIAC	Electronic Numerical Integrator and Computer
ESO	European Southern Observatory
EUV	Extreme ultraviolet
EUVE	*Extreme Ultraviolet Explorer*

H.M. Van Horn, *Unlocking the Secrets of White Dwarf Stars*,
Astronomers' Universe, DOI 10.1007/978-3-319-09369-7

EXOSAT	*European X-ray Observatory Satellite*
eV	Electron volt
FFT	Fast Fourier Transform computer algorithm
FORTRAN	FORmula TRANslation computer language
FOS	*Faint Object Spectrograph*
FUSE	*Far Ultraviolet Spectroscopic Explorer*
FUV	Far ultraviolet
GHRS	*Goddard High-Resolution Spectrograph*
Gpc	Gigaparsec = one billion parsecs
Gyr	Gigayear = one billion years
H	Hydrogen
He	Helium
HE	Hamburg-ESO survey
HEAO	*High Energy Astrophysics Observatory*
H/He/C	Hydrogen, helium, and carbon layers
H-R	Hertzsprung-Russell (diagram)
HS	Hamburg-Schmidt survey
HST	*Hubble Space Telescope*
Hz	Hertz, a unit of frequency
HZ	Humason-Zwicky star
IAU	International Astronomical Union
IUE	*International Ultraviolet Explorer*
IBM	International Business Machines company
JILA	Joint Institute for Laboratory Astrophysics
K	(degrees) Kelvin (measure of absolute temperature)
keV	Kilo-electron volt (=1,000 eV)
kG	Kilogauss = 1,000 G (measure of magnetic field strength)
kpc	Kiloparsec = 1,000 pc
ΛCDM	Lambda Cold Dark Matter cosmological model
LANL	Los Alamos National Laboratory
LB	Luyten Blue survey
LESIA	Laboratoire d'Etude Spatiale et d'Instrumentation en Astrophysique
LHS	Luyten Half Second catalog
LIGO	Laser Interferometer Gravitational-Wave Observatory
LLNL	Lawrence Livermore National Laboratory
log g	Logarithm of surface gravity

LP	Luyten Palomar survey
LPL	Lunar and Planetary Laboratory
LSU	Louisiana State University
LTE	Local thermodynamic equilibrium
LTT	Luyten Two-Tenths catalog
MCT	Montreal-Cambridge-Tololo survey
MeV	Million electron volt
MG	Megagauss = 1 million Gauss (measure of magnetic field strength)
MIT	Massachusetts Institute of Technology
MMT	Multi-Mirror Telescope
Mpc	Megaparsec = one million parsecs
MWG	Milky Way Galaxy
NASA	National Aeronautics and Space Administration
NLTT	New Luyten Two-Tenths catalog
OAO	Orbiting Astronomical Observatory
OCP	One-Component Plasma
ONe	Oxygen-neon mixture
OP	Opacity Project
OPAL	Opacity Project at LLNL
OSO	Orbiting Solar Observatory
pc	Parsec
PG	Palomar Green survey
PNN	Planetary nebula nucleus
PNNV	Variable planetary nebula nucleus
POSS1	Palomar Objective Sky Survey, *ca.* 1950
RA	Right ascension
RGB	Red giant branch
ROSAT	*Röntgen SATellite*
sdB	Subdwarf B star
sdBV	Variable subdwarf B star
sdO	Subdwarf O star
SDSS	Sloan Digital Sky Survey
SN	Supernova
SNe	Supernovae
SN Ia	Type Ia supernova
STIS	*Space Telescope Imaging Spectrograph*
STS	Space Transportation System
UBVRI	Ultraviolet, blue, visible, red, infrared photometric colors

UCI	University of California at Irvine
UCLA	University of California at Los Angeles
UNIVAC	UNIVersal Automatic Computer
USNO	US Naval Observatory
UV	Ultraviolet
uvby	Ultraviolet, violet, blue, yellow Strömgren photometric colors
WDLF	White dwarf luminosity function
WET	Whole Earth Telescope
XCOV	Extended coverage
ZAMS	Zero-age Main Sequence

Symbols

3α	Triple-alpha He-burning reaction
Å	Ångstrom unit (10^{-8} cm)
c	Speed of light
d	Distance
E	Energy
F	Free energy (a thermodynamic quantity)
γ	Symbol for a photon
Γ	Ratio of electrostatic to thermal energy
g	Surface gravity (of a star)
h	Planck's constant
$\hbar$	Planck's constant divided by 2π
Hα	First Balmer line of hydrogen, at 6,563 Å
Hβ	Second Balmer line of hydrogen, at 4,861 Å
Hγ	Third Balmer line of hydrogen, at 4,340 Å
k, l, m	Pulsation-mode indices
λ	Wavelength
L	Luminosity (of a star)
L_{Sun}	Luminosity of Sun
m	Mass of a subatomic particle
M_{bol}	Absolute bolometric magnitude
M	Mass (of a star)
M_{Sun}	Mass of Sun
m_V	Apparent visual magnitude
M_V	Absolute visual magnitude

ν	Frequency
$n, l, m,$	Quantum numbers of atomic energy levels
n(H)	Number density of hydrogen atoms
n(He)	Number density of helium atoms
Ω	Rotation rate (of a star); also cosmological mass density
p	Momentum of a subatomic particle
P_{rot}	Rotation period (of a star)
R	Radius (of a star)
R_{Sun}	Radius of Sun
T_{eff}	Effective temperature
Z	Atomic number

Glossary

Absolute magnitude The apparent magnitude a star would have if placed at a standard distance of 10 pc from the Sun.
Absolute bolometric magnitude (M_{bol}) The absolute magnitude measured across all wavelengths. It is proportional to the logarithm of the total luminosity of the star.
Absolute visual magnitude (M_V) The apparent visual magnitude a star would have if placed at a standard distance of 10 pc.
Absolute temperature The temperature measured from absolute zero in degrees Kelvin (also called Kelvins, abbreviated as "K").
Absolute zero The temperature at which there is a complete absence of any thermal energy. It occurs at about 273° below 0 on the Celsius, or Centigrade, scale. (0 °C is set at the freezing point of water –32 °F—and 100 °C is set at the boiling point of water –212 °F.)
Abundance In reference to a chemical element, the fraction of the total mass (or number of atomic nuclei) in a fixed quantity of matter that is contributed by atoms or ions of a given chemical element.
Accretion disk A gaseous disk formed in a close binary system by mass transfer from a Roche-lobe-filling companion.
Accretion rate The rate at which matter is accreted onto a star from a companion or from the interstellar medium.
Achromatic lens A composite lens designed so that light of different colors comes to a focus at the same point.
Adiabatic change A change that is not accompanied by heat flows.
Alias A spurious frequency produced by the "beating" of an actual stellar oscillation frequency with a non-stellar frequency such as the frequency associated with Earth's daily rotation.
AM Canum Venaticorum A close binary cataclysmic variable in which a low-mass, degenerate helium star is transferring mass to its white dwarf companion.
Ambient The local conditions in the vicinity of a given point.

H.M. Van Horn, *Unlocking the Secrets of White Dwarf Stars*,
Astronomers' Universe, DOI 10.1007/978-3-319-09369-7

Amplitude-variable oscillation An oscillation in which the amplitude is not fixed, as it is in a sinusoid, but which varies during the course of a measurement.

Ångstrom (Å) A unit of distance equal to one hundred-millionth of a centimeter, or about the size of an atom.

Angular momentum A physical property of a spinning body that remains constant unless acted upon by an external torque. It has the dimensions of mass times radius times rotation rate. Conservation of angular momentum is the reason why a skater's rate of spin increases as she pulls her arms in close to her body, decreasing her effective radius.

Apastron The point in its elliptical orbit at which a body is farthest from its parent star.

Aphelion The point in its elliptical orbit at which a planet is farthest from the Sun.

Apparent magnitude A logarithmic measure of the brightness of a star. The original determinations of the relative brightness of stars were made by eye in antiquity. The current definition, adopted in 1850 with subsequent refinements, defines five magnitudes to correspond to a difference in apparent brightness of a factor of 100. Since the logarithm of 100 is 2.0, a difference of one magnitude corresponds to a difference in the logarithm of 0.40, or to a brightness ratio of a factor of 2.512. The scale is fixed so that stars from the second to the fifth magnitude have the traditional values. On this scale, Sirius A has an apparent magnitude of −1.46.

Apparent bolometric magnitude The apparent magnitude measured across all wavelengths.

Apparent visual magnitude The apparent magnitude measured in the visual wavelength band, a range of wavelengths about 900 Å wide centered on 5,500 Å.

Arc second One second of arc.

Asteroseismology The use of observed stellar pulsation frequencies or periods combined with calculations of theoretical stellar models to determine the physical properties of a star.

Astronomical unit (AU) The distance from Earth to the Sun, approximately 150 million km. Often used as a unit of measure for other Solar System objects.

Asymptote A curve that gradually approaches a limiting value as a parameter that varies along the curve becomes arbitrarily large.

Asymptotic giant branch (AGB) A track in the Hertzsprung-Russell diagram that approximately parallels the red giant branch (RGB) but at higher luminosity.

Atmospheric extinction The dimming of light from an astronomical object that is produced by Earth's atmosphere.

Black body radiation The thermal radiation emitted from a perfectly absorbing body, or from a small hole in a cavity in thermal equilibrium.

Black hole An astrophysical body that is so dense that not even light can escape from it. Einstein's general theory of relativity is needed to understand it.

Blue edge The short-wavelength limit of a spectral region.

Bolometric correction The difference between the absolute bolometric magnitude, which is related to total power output of a star, and the absolute visual magnitude.

Calculus A branch of mathematics that deals with infinitesimal changes.

Cataclysmic variable A close binary system containing a white dwarf and a companion star that undergoes episodic outbursts of energy.

CCD detectors Solid-state charge-coupled devices that are sensitive, two-dimensional detectors of photons.

Celestial coordinates The celestial latitude and longitude of an astronomical object, which locate it precisely in the sky.

Celestial sphere An imaginary sphere in space centered on Earth. The celestial poles are the projections of Earth's North and South poles onto the celestial sphere, and the celestial equator is the projection of Earth's equator. Angular measures of celestial latitude are termed "declination," and angular measures of celestial longitude are termed "right ascension."

Celestial equator The projection of Earth's equator onto the celestial sphere.

Center of mass The point in space that moves as if all the mass of the system is concentrated there. For a binary system such as Sirius A and B, the center of mass lies on a line connecting the centers of the two stars. The distance of each star from the center of mass is inversely proportional to its own mass. The lower-mass star thus exhibits a proportionately larger range of motion than the larger one.

Central stars of planetary nebulae (CSPN) Also called planetary nebula nuclei or PNNs, these are hot, degenerate stars found in the centers of the glowing masses of gas called planetary nebulae. The strong ultraviolet radiation from the central star keeps the nebula ionized.

Cepheid variable A member of a class of massive Main Sequence stars that undergo spontaneous radial pulsations.

CGS system of units A consistent system of physical units based on length measured in centimeters, mass measured in grams, and time measured in seconds.

Circular polarization An electromagnetic wave, such as visible light or a radio wave, can be thought of as consisting of a time-varying electric field—represented as a vector pointing in a given direction—and a time-varying magnetic field, both field directions perpendicular to the direction of propagation of the wave. If the wave is propagating along the *z*-axis, the two field vectors lie in the *x*–*y* plane. The two fields oscillate out of phase with each other, one field increasing in strength as the other decreases, the frequency of the electromagnetic wave being the frequency of oscillation. When the vector representing the electric field makes a complete circuit around the direction of propagation during each cycle of the wave—like the minute hand circling around the face of a clock every hour—the wave is said to be circularly polarized.

CNO elements The chemical elements carbon, nitrogen, and oxygen, which are often found together.

Collodion A nitrocellulose coating that once was used to coat glass plates for photography.

Color magnitude diagram A graph showing the (absolute or apparent) magnitudes of stars plotted against their photometric colors.

Comet A primordial mass of rock and ice that normally exists in the far outer reaches of space. When a comet in our Solar System is drawn in toward the Sun, heating by sunlight evaporates matter from the cometary nucleus, producing spectacular, long tails that extend far across the sky.

Common envelope phase A stage in the evolution of a close binary system in which one star expands so rapidly that both stars are engulfed in its envelope.

Compiler A program that transforms a computer code written in a programming language into a set of instruction in "machine language" that can be executed by a computer.

Composition-transition zone A region in a star in which the mixture of chemical elements changes smoothly from one composition to another.

Continuum The continuous spectrum of a star.

Convective driving A mechanism that operates in a stellar convection zone to excite oscillations in a star.

Convective mixing The stirring produced by fluid motions associated with convection.

Convection zone A region in a star in which energy is transported by convection.

CoRoT A European spacecraft designed to study *Convection, Rotation, and planetary Transients*.

Cosmic Background Explorer (COBE) A NASA spacecraft launched in 1989 to make precise measurements of the cosmic microwave background radiation.

Cosmic microwave background (CMB) Thermal radiation remaining from the Big Bang, which now has a temperature of about 3 K, corresponding to electromagnetic radiation in the microwave part of the spectrum.

Cosmological constant A term in Einstein's general relativistic equations of gravitation that affects the expansion rate of the Universe. Originally introduced by Einstein before the Universe was known to be expanding, subsequently regarded by him as his "greatest blunder," but recently resurrected to account for the accelerating expansion of the Universe.

Cyclotron frequency The frequency at which an electron moves in a circular orbit in a strong magnetic field. The name comes from a machine called a cyclotron that was first designed to accelerate protons to high energies in circular magnetic orbits.

DA white dwarf A white dwarf with a pure hydrogen spectrum.

DAB white dwarf A white dwarf with a predominantly hydrogen spectrum but which also contains traces of neutral helium.

DAO white dwarf A white dwarf with a predominantly hydrogen spectrum but which also contains traces of ionized helium.

Daughter mode One of two oscillation modes into which an original "parent" mode can transfer energy by means of nonlinear coupling mechanisms.

DAV star A pulsating DA white dwarf, also known as a ZZ Ceti star.

DAZ white dwarf A white dwarf with a predominantly hydrogen spectrum but which also contains lines of metallic elements.

DB gap The range of effective temperature from about 45,000 K to 30,000 K in which no DB white dwarfs are detected.

DB white dwarf A white dwarf with a pure neutral helium spectrum.

DBA white dwarf A white dwarf with a predominantly neutral helium spectrum but which also contains traces of hydrogen.

DBV star A pulsating DB white dwarf, also known as a V477 Her star.

DC white dwarf A white dwarf with no lines in its spectrum.

DO white dwarf A white dwarf with a pure ionized helium spectrum.

DOV star A pulsating DO white dwarf, also known as a GW Vir or PG1159 star.

DQ white dwarf A white dwarf with traces of carbon in its spectrum.

DZ white dwarf A white dwarf with lines of metallic elements in its spectrum.

Dark energy Energy that causes the expansion of the Universe to accelerate. The nature of this mysterious energy is currently unknown.

Dark matter Matter that is detected only through its gravitational influence on surrounding visible matter. The nature of dark matter is currently unknown.

Debye temperature A characteristic temperature below which the heat capacity of a solid drops rapidly to zero.

Declination The celestial latitude of an astronomical object north (positive angles) or south (negative angles) from the celestial equator. Declination is measured in degrees, minutes, and seconds of arc.

Degenerate The condition in which electrons fill up all the lowest available energy states in a system at low temperature.

Differential equation An equation involving the derivatives (as defined in calculus) of one or more quantities.

Differential rotation Rotation of a star in which different layers rotate at different rates.

Diffusion The gradual spreading out of the particles of a chemical species—atoms, ions, or molecules—from a region of higher concentration to one of lower concentration.

Diffusion coefficients The ratio of the rate of diffusion to the concentration gradient.

Diurnal Daily.

Doppler shift The shift in wavelength produced by the motion of a source of radiation. Motion away from the observer causes a lengthening of the wavelength (redshift), while motion toward the observer shortens the wavelength (blueshift).

Dwarf nova An erupting binary system in which a white dwarf accretes matter from a non-degenerate companion.

Effective temperature (T_{eff}) Proportional to the one-fourth root of the luminosity of a star divided by the square root of its radius, the effective temperature can be thought of as the surface temperature of a star.

Electron conduction The process by which heat is efficiently transported along a temperature gradient in a metal or in the electron-degenerate interior of a star.

Emergent flux or intensity The flux or intensity of radiant energy emitted from the surface of a star.

Energy density The energy contained in a unit volume. In the CGS system of units, it is measured in ergs per cubic centimeter.

Entropy A thermodynamic quantity that measures the heat content of a system.

Erg The unit of energy in the centimeter-gram-second (CGS) system of physical units.

EUVE NASA's *Extreme Ultraviolet Explorer* spacecraft launched in 1992 to permit astronomical observations in the wavelength range from 100 to 700 Å, which cannot be studied from the ground.

Excitation The process of raising an electron in an atom or ion from one quantum state to a quantum state with higher energy. Also, a process that produces spontaneous oscillations in stars.

Exponent A mathematical quantity. In the equation $y = x^n$, the exponent is the quantity n.

Far UV or FUV The short-wavelength portion of the ultraviolet spectrum, from about 900 to 1,180 Å.

Fast Fourier Transform (or FFT) A computer algorithm that efficiently decomposes a time series of measurements of fixed duration into components with different frequencies. If applied to a tone from an organ pipe, for example, the FFT yields the fundamental and all the harmonics present in the tone.

Filter transmission function The wavelength-dependent transparency of a filter material in a given wavelength band.

"Fingers of God" Columns of interstellar matter in the Eagle Nebula, as imaged by the *Hubble Space Telescope*.

FORTRAN An early computer programming language that enabled easy construction of programs containing complicated mathematical operations. The name is a shortened version of "formula translation."

g-modes Stellar oscillations for which the restoring force is gravity acting on the thermally induced buoyancy of mass elements.

Gamma mechanism (or γ mechanism) A mechanism that operates in a stellar partial ionization zone to excite pulsations in the star.

General relativity Einstein's theory of gravitation as a consequence of the warping of spacetime by a massive body.

Gigayear One billion (or 1,000 million) years.

Gigaparsec One billion parsecs.

Globular cluster An approximately spherical cluster of perhaps a 100,000 stars. Hundreds of globular clusters orbit at large distances around the center of the Milky Way Galaxy. The stars in a globular cluster are among the oldest in the Galaxy.

Gravitational radiation The emission of energy from a system by wavelike disturbances in the spacetime continuum. This is one of the effects predicted by Einstein's general-relativistic theory of gravitation.

Gravitational redshift The shift in wavelength of a spectral line emitted from the surface of a massive body as measured at large distances relative to the laboratory wavelength of the line. The gravitational redshift is a prediction of Einstein's general relativistic theory of gravity.

Gravitational segregation The sorting of chemical elements under gravity, with the lighter elements rising and the heavier ones falling.

"Great Attractor" An enormous concentration of matter equivalent to many tens of thousands of galaxies—most likely a supercluster of galaxies—lying some 80 Mpc away.

GW Virginis star A very hot, degenerate, pulsating star such as PG 1159-035.

Halley's Comet The comet for which Edmund Halley published the orbit in 1705, showing that it returns close to the Sun every 76 years. Its most recent perihelion passage occurred in 1986.

Hertzsprung-Russell diagram (H-R diagram) A chart showing the luminosities (or absolute magnitudes) of stars plotted against their effective temperatures (or colors).

Hipparcos A European astrometry spacecraft launched in 1989.

Hot subdwarf Subdwarf stars lie between the Main Sequence and white dwarfs in the H-R diagram. Hot subdwarfs of spectral types B and O are thought to be in advanced stages of stellar evolution on their way to becoming white dwarfs.

Hubble constant The constant of proportionality in Hubble's law.

Hubble's law A linear relationship between the distance to a far Galaxy and the redshift of absorption lines in its spectrum.

Hubble Space Telescope A 2.4-m telescope carried into low Earth orbit in 1990 and operated by the Space Telescope Science Institute for a wide range of imaging, photometric, and spectroscopic investigations.

Hydrostatic equilibrium The physical state of a fluid body in which all forces are balanced so that there are no fluid motions.

IAU The International Astronomical Union.

IBM cards Standardized 80-column cards on which data or computer programs were encoded by means of sequences of punched holes in successive columns. Used at one time by the IBM company

Inner Lagrangian point The point midway between two gravitating bodies at which the gravitational force of one is exactly balanced by the gravitational force of the other. Named for the famed French scientist Lagrange.

Interstellar matter Although the space between the stars contains less matter than even the best vacuum that can be produced in a terrestrial laboratory, it is not completely empty. Instead it contains atoms, ions and electrons, molecules, and even solid particles, albeit at extremely low density. This matter can be seen as concentrations of gas and dust, especially around young stars, which form inside such clouds.

Ionization The process of removing one or more electrons from a neutral atom or an electrically charged ion.

IUE The *International Ultraviolet Explorer* spacecraft, launched in 1978, that has been used for a wide range of spectroscopic investigations in the ultraviolet.

Kappa mechanism (or κ mechanism) A mechanism that operates in a stellar partialionization zone to excite pulsations in the star.

Kilogauss (kG) A magnetic field strength of 1,000 G.

Kiloparsec (kpc) A distance of 1,000 pc.

Landau levels Circular electron orbits in a strong magnetic field.

Latent heat The change in energy associated with a first-order phase change. A familiar example is provided by the energy needed to melt ice into water with no change in the temperature. This is the reason ice is able to keep a drink cold as it gradually melts.

LIGO The Laser Interferometer Gravitational-wave Observatory, a facility constructed to detect the very weak disturbances in spacetime produced by gravitational waves.

Line blanketing The effect of absorption lines upon the stellar continuum emission. It consists of "line blocking," wherein flux is removed by the absorption lines, and "backwarming," in which the level of the continuum emission between the dark lines is increased, as if the star were slightly hotter than had been assumed.

Linear polarization Like circular polarization, except that the electric vector always points in the same fixed direction as the amplitude of the field varies around each cycle of the wave. Also called "plane polarization."

Local Group A group of galaxies, including the Milky Way and the Andromeda Galaxy, which are gravitationally bound to each other.

Local thermodynamic equilibrium (LTE) A condition in which the gradients of thermodynamic quantities such as pressure and temperature across a given parcel of matter in a star are so small that the parcel can be considered to be almost in complete thermodynamic equilibrium.

Luminosity The intrinsic power output of a star. The luminosity of the Sun is approximately 4×10^{33} ergs/s $= 4 \times 10^{26}$ J s^{-1} $= 4 \times 10^{26}$ W. The absolute bolometric magnitude of a star is proportional to the logarithm of its luminosity: $M_{bol} = 4.72 - 2.5 \log(L/L_{Sun})$, where the absolute bolometric magnitude of the Sun is +4.72, and L is the luminosity of the star.

Magnitude A logarithmic measure of the brightness of a star.

Main Sequence The band of stars in the H-R diagram that are in the process of central hydrogen burning.

Mass-radius relation for white dwarfs The relation between the mass M and radius R for theoretical zero-temperature models of white dwarfs.

Megagauss (MG) A magnetic field strength of one million gauss.

Megaparsec (Mpc) A distance of one million parsecs.

Meridian A circle of celestial longitude that passes through the celestial poles and the local zenith.

Meridian circle A device to enable the precise measurement of the time and polar angle of a star passing the local meridian.

Micrometer A device that enables the accurate measurement of small distances.

Milky Way Galaxy Our home Galaxy, thought to be a spiral Galaxy. The Solar System lies about 8,500 pc out from the galactic center.

Multi-periodic oscillation A pulsation that displays several different oscillation modes simultaneously.

Non-DA white dwarfs White dwarfs that have no H lines in their spectra.

Nonlinear energy transfer A process that can transfer energy between oscillation modes.

Nova An erupting star in which hydrogen-rich matter that has been accreted onto the surface of a white dwarf undergoes an explosive thermonuclear runaway.

Nucleogenesis The creation of elements by means of nuclear reactions.

Nucleosynthesis The process of synthesizing elements by means of nuclear reactions.

Objective lens The lens at the front of a refracting telescope that brings the parallel rays of light from a distant object to a focus.

Objective prism A glass plate ground to a slight angle—slightly thinner on one side than the other—that is placed at the front of a telescope. The slightly tapered glass plate forms a prism that disperses the light from a star according to its wavelength. Instead of being a round dot of light, each star thus appears in the resulting image as a tiny rainbow spectrum.

Ohmic dissipation The draining of electric or magnetic energy caused by the electrical resistance of matter. The unit of electrical resistance is the ohm, which gives this process its name.

Opacity A physical quantity that measures the resistance of a parcel of matter to a flow of radiant energy. It includes the effects of both absorption and scattering.

Open cluster An irregular star cluster in the galactic disk that contains scores to thousands of stars.

Oscillation spectrum The collection of independent oscillation modes present, *e.g.*, in a pulsating star.

***p*-modes** Stellar oscillations for which the restoring force is the variation of pressure with density.

Parallax The angular displacement of a nearby star against the background of distant stars that is caused by the motion of Earth around the Sun during the year. The effect is the same as the apparent displacement of a nearby object as seen against more distant scenery that is caused by viewing it first with one eye open (and the other closed) and then with the other eye open (and the first eye closed).

Parsec (pc) The distance from the Sun at which an astronomical object has a parallax of exactly one second of arc. This distance is $(360°/2\pi)\times(60'/1°)\times(60''/1')$ AU = 206,265 AU.

Partial ionization zone A region in which matter is partially ionized.

Periastron The point in its elliptical orbit at which a body is closest to its parent star.

Perihelion The point in its elliptical orbit at which a planet is closest to the Sun.

PG 1159-035 A very hot, degenerate star, similar to the central star of a planetary nebula. Originally discovered by astronomer Richard Green in a survey for stellar objects that radiate strongly in the ultraviolet. This star was subsequently found to be a variable star.

Photometer A device that enables measurements of the intensity of starlight.

Photometry The measurement of portions of a star's spectrum in different wavelength bands.

Photon The quantum of electromagnetic radiation, a "particle" of light.

Photosphere The visible surface of a star.

Planetary nebula A highly ionized mass of hot gas surrounding an extremely hot, small, dense star. The nebulosity was ejected by the star through a process that is still not well understood. The central star of a planetary nebula is the stellar remnant left at the end of the AGB phase of evolution and is on its way to becoming a white dwarf.

Planetary nebula nucleus (PNN) See "Central stars of planetary nebulae".

Plasma An overall neutral gaseous mixture of positively charged ions and negatively charged electrons.

PNNV A planetary nebula nucleus (the central star of a planetary nebula) that is a variable star.

Polar A close binary system containing a strongly magnetic white dwarf.

Polytrope A simplified, spherically symmetric model for a star in which the pressure is given by a polytropic equation of state.

Polytropic equation of state An expression giving the local pressure as a power of the local density.

Power spectrum A graph that shows a distribution of peaks at different oscillation frequencies, with the tallest peaks representing the most prominent periodicities in the spectrum.

Precession The change in direction of the axis of rotation of a spinning body that is produced by an external torque. A familiar example is the 26,000-year precession of Earth's rotation axis produced by the gravitational pull of the Sun and Moon on Earth's equatorial bulge.

Pressure ionization An effect of the fluctuating electric fields in a stellar atmosphere that lowers the ionization potential of an absorbing atom. This comes about because the electrical charges of nearby ions and electrons weaken the electrostatic potential by which electrons are bound to an atomic nucleus, thereby making it possible for photons of lower energy to ionize the atom. This effect depends upon the local average density of charged particles, with a greater reduction in the ionization potential being produced by higher densities.

Proper motion The angular motion of a star or other astronomical body across the celestial sphere. A star with a large proper motion tends to be relatively close to the Sun, while one with a smaller proper motion tends to be farther away.

Pulsation damping mechanism A physical process that acts to extract energy from an oscillation mode. Without a compensating process to add energy into the mode, damping will cause the amplitude of the mode to decay to zero.

Pulsation excitation mechanism A physical process that pumps energy into a given mode of oscillation.

Pulsational instability strip A narrow region of the H-R diagram in which a particular class of star spontaneously develops pulsations.

Pycnonuclear reactions Nuclear reactions induced in matter at very high density.

Radial oscillation An oscillation of a spherical star that involves only inward and outward motions, with no transverse motions.

Radiation pressure Photons of electromagnetic radiation produce a definite pressure. This pressure becomes increasingly important at very high temperatures, such as occur in the interiors of stars.

Random walk Also called a "drunkard's walk," this is a process by which a particle moves from one point in space to another by taking a sequence of steps, each following in an arbitrary direction from the one before.

Red edge The long-wavelength limit of a spectral region.

Red giant branch (RGB) A trajectory in the H-R diagram along which stars ascend to large radii and high luminosities while the core contracts toward the ignition of central helium burning.

Redshift An increase in the wavelength of a spectral line. The increase may be caused by the recession velocity of the object emitting the line (Doppler shift) or by the source lying deep in a gravitational potential well (gravitational redshift or Einstein shift).

Refraction The bending of a light ray when it crosses a boundary between two different media, as for example, the boundary between air and water or between air and the glass of a prism or lens.

Relativistic Effects that occur when the speed of an object approaches the speed of light.

Reticle Cross hairs mounted in the eyepiece of a telescope to enable precise positioning on an astronomical object.

Right ascension The celestial longitude of an astronomical object measured from the vernal equinox, the unique point on the celestial sphere where the Sun's apparent path crosses the celestial equator heading north. Right ascension is measured in hours, minutes, and seconds of time.

ROSAT The European Röntgen Satellite, an imaging X-ray telescope launched in 1990.

Roche lobe The largest region in a close binary system in which matter can be considered to be gravitationally bound to one of the two stars. It is a teardrop-shaped region of space, with the pointed end directed toward the companion star.

Roche model A simplified model of a close binary system that contains two point masses in circular orbits around each other.

Rosseland opacity An approximate expression for the opacity of stellar matter originally developed by Norwegian astrophysicist Svein Rosseland.

sdBV A hot subdwarf of spectral type B that exhibits pulsations.

Sforzando event An abrupt, dramatic change in the power spectrum of a pulsating white dwarf.

Shell flash An event produced by the rapid burning of fuel in a nuclear-burning shell source in an evolved star.

Shoemaker-Levy 9 A comet that impacted the planet Jupiter in 1994.

Shot noise Noise resulting from the detection of insufficient numbers of photons.

Sine curve (sinusoid) A regular oscillation with a fixed wavelength and fixed maxima and minima of amplitude.

Sloan Digital Sky Survey (SDSS) A project that, beginning in 2000, has been conducting deep surveys of about one-quarter of the entire sky, obtaining digital images in multiple wavelength bands. It has proven useful for a wide range of astronomical studies of very faint objects.

Solar luminosity (L_{Sun}) The total power output of the Sun, about 4×10^{26} W.

Solar mass (M_{Sun}) The mass of the Sun, about 2×10^{30} kg.

Spectral classification The grouping of stellar spectra based on the appearance of spectral lines produced by various elements in different stages of ionization and excitation.

Spectral type A label assigned to a stellar spectrum based on its appearance. For Main Sequence stars, the spectral types are O, B, A, F, G, K, M, ranging from effective temperatures exceeding 40,000 K down to those below 2,800 K. White dwarf stars have their own, separate spectral classification system, described in the text.

Spectropolarimetry Polarization measurements carried out at different wavelengths.

Spectroscope An instrument that separates light into the different wavelengths present.

Spectroscopist An observational astronomer who specializes in spectroscopy.

Spectrum Visible light or other electromagnetic waves dispersed in order of wavelength.

Spherical harmonic A function that describes the "wiggliness" of a quantity such as a stellar oscillation in latitude and longitude.

Standard candle An object with a known intrinsic brightness. The distance to the object can be determined by comparing its apparent brightness to its known intrinsic brightness.

Stefan-Boltzmann law The physical law stating that the total power radiated by a black body is proportional to the fourth power of its temperature.

Supernova An exploding star that may temporarily be as bright as all the stars in an entire Galaxy.

Surface gravity The acceleration, *g*, due to gravity at the surface of a massive body. For white dwarfs it is often convenient to use the logarithm of this quantity, log *g*.

Thermonuclear reactions Nuclear reactions induced by very high temperatures.

Time series A series of measurements taken continuously, one after another, at discrete instants in time.

Trapped mode A pulsation mode in which the oscillatory motions are confined to a limited region of the star.

UBV photometry Logarithmic measurements of the flux of light from a star taken through broad-band filters centered on 3,600 Å (U), 4,400 Å (B), and 5,500 Å (V). The system is calibrated so that the V measurement gives the apparent visual magnitude of the star. Because the system is logarithmic, differences—like the B-V color index—cancel out effects such as an unknown distance and are useful intrinsic properties of the star related to its effective temperature and composition.

Universal Time (UT) is a modern version of mean solar time at the Prime Meridian in Greenwich, England.

UV spectrum The portion of the electromagnetic spectrum at ultraviolet wavelengths, short of the cutoff produced by Earth's atmosphere at about 3,200 Å down to about 100 Å.

uvby photometry Also called "Strömgren photometry" after its originator, this system is based on narrowband measurements through filters centered on 3,500 Å (u), 4,100 Å (v), 4,700 Å (b), and 5,550 Å (y). Similarly to broadband photometry, the color index $(b-y)$ provides a measure of the star's effective temperature. The filter wavelengths were carefully chosen, however, so that the index $c_1 = (u-v) - (v-b)$ is sensitive to the surface gravity of the star, while the index $m_1 = (v-b) - (b-y)$ provides a measure of the stellar "metallicity," or heavy-element content.

V477 Herculis The prototype variable DB (He-atmosphere) white dwarf. Also called GD 358 or WD1645+325.

Viscous dissipation The conversion of the energy in a shear flow into heat by the action of viscosity.

White dwarf luminosity function (WDLF) A logarithmic plot of the numbers of white dwarfs in successive intervals of luminosity. Because white dwarfs cool increasingly slowly as they become fainter, the numbers of these stars increase toward fainter luminosities.

Whole Earth Telescope (WET) A collection of observatories spread around Earth in longitude that was organized to enable coordinated observations of a star.

XCOV An "extended coverage" series of observations of a star conducted with the Whole Earth Telescope.

Zeeman splitting The separation caused by a strong magnetic field of the components of a spectral line having different quantum numbers.

ZZ Ceti star A pulsating DA white dwarf.

Bibliography

Beauchamp, A.: Ph.D. Thesis, Université de Montréal, Détermination des Paramétrés Atmosphériques des Étoiles Naines Blanches de Type DB (1995)

Beer, A. (ed.): Vistas in Astronomy, vol. 15. Pergamon, Oxford (1973)

Bergeron, P.: Ph.D. Thesis, Université de Montréal, Propriétés Atmosphériques des Étoiles Naines Blanches Froides de Type DA (1988)

Bethe, H.A.: The Road from Los Alamos. American Institute of Physics, New York (1991)

Chandrasekhar, S.: An Introduction to the Study of Stellar Structure. University of Chicago, Chicago (1939); Reprinted in 1957 by Dover

DeVorkin, D.H. (ed.): The American Astronomical Society's First Century. American Astronomical Society, Washington, DC (1999)

Eddington, A.S.: The Internal Constitution of the Stars. Cambridge University Press, Cambridge (1926)

Feynman, R.P.: Surely You're Joking, Mr. Feynman! Bantam Books, Toronto (1986)

Fontaine, G.: Ph.D. Thesis, University of Rochester, Outer Layers of White Dwarf Stars (1973)

French, A.P.: Principles of Modern Physics. Wiley, New York (1958)

Gamow, G.: The Birth and Death of the Sun. Penguin, New York (1945)

Green, R.F.: Ph.D. Thesis, Caltech, A Complete Sample of White Dwarfs, Hot Subdwarfs, and Quasars (1977)

Hamada, T., Salpeter, E.E.: Astrophs. J., 134, 669, "Models for Zero Temperature Stars." (1961)

Hansen, C.J., Kawaler, S.D.: *Stellar* Interiors. Springer, New York (1994)

Holberg, J.B.: Sirius: Brightest Diamond in the Night Sky. Springer, New York (2007)

Hunger, K., Schönberner, D. (eds.): IAU Colloq. No. 87: Hydrogen Deficient Stars and Related Objects. Reidel, Dordrecht (1986)

Kangro, H.: Planck's Original Papers in Quantum Physics. Taylor & Francis, London (1972)

Kawaler, S.D.: Ph.D. thesis, University of Texas (Austin), The Transformation from Planetary Nebula Nucleus to White Dwarf: A Seismological Study of Stellar Metamorphosis (1986)

Kittel, C., Kroemer, H.: Thermal Physics W. H. Freeman, San Francisco (1980)

H.M. Van Horn, *Unlocking the Secrets of White Dwarf Stars*,
Astronomers' Universe, DOI 10.1007/978-3-319-09369-7

Koester, D.: Ph.D. Thesis, Kiel University, Äussere Hüllen und die Abkühlung von Weissen Zwerge, Outer Layers and the Cooling of White Dwarfs (1971)

Kondo, Y., et al.: Exploring the Universe with the IUE Satellite. Reidel, Dordrecht (1987)

Kopal, Z.: Close Binary Systems. Chapman & Hall, London (1959)

Lamb, D. Q., Jr.: Ph.D. Thesis, University of Rochester, Evolution of Pure ^{12}C White Dwarfs (1974)

Langer, W.L.: An Encyclopedia of World History. Houghton Mifflin, Boston (1968)

Logsdon, J.M.: Exploring the Unknown. NASA History Office, Washington, DC (2001)

Luyten, W.J. (ed.): White Dwarfs. Reidel, Dordrecht (1971)

Marshak, R.E.: Academic Renewal in the 1970s: Memoirs of a City College President. University Press of America, Washington, DC (1982)

Melissinos, A.C., Van Horn, H.M. (eds.): Robert E. Marshak, 1916—1992, Tributes to His Memory. University of Rochester, Rochester (1993)

Misner, C.W., Thorne, K.S., Wheeler, J.A.: Gravitation. W. H. Freeman, San Francisco (1973)

Pais, A.: 'Subtle is the Lord…', The Science and the Life of Albert Einstein. Clarendon, Oxford (1982)

Pannekoek, A.: A History of Astronomy. Dover, New York (1989)

Philip, A.G.D., Hayes, D.S., Liebert, J.W. (eds.): IAU Colloq. No. 95: Second Conference on Faint Blue Stars. L. Davis, Schenectady (1987)

Rhodes, R.: The Making of the Atomic Bomb. Simon & Schuster, New York (1986)

Rhodes, R.: Dark Sun, the Making of the Hydrogen Bomb. Simon & Schuster, New York (1995)

Schatzman, E.: White Dwarfs. North Holland, Amsterdam (1958)

Schwarzschild, M.: Structure and Evolution of the Stars. Princeton University Press, Princeton (1958)

Seaton, M.J.: The Opacity Project. Institute of Physics, Bristol (1995)

Shapiro, S.L., Teukolsky, S.A.: Black Holes, White Dwarfs, and Neutron Stars. Wiley, New York (1983)

H.L. Shipman, Ph.D. Thesis, Caltech, White Dwarfs (1971)

Smiley, J.: The Man Who Invented the Computer. Doubleday, New York (2010)

Soifer, B.T. (ed.): Sky Surveys: Protostars to Protogalaxies. ASP Conf. Ser, vol. 43. Astronomical Society of the Pacific, San Francisco (1993)

Solheim, J.-E., Meištas, E.G. (eds.): 11th European Workshop on White Dwarfs. ASP Conf. Ser, vol. 169. Astronomical Society of the Pacific, San Francisco (1999)

Spitzer, L.: Astronomical Advantages of an Extra-Terrestrial Observatory (Project RAND: Santa Monica (1946)

Unno, W., Osaki, Y., Ando, H., Shibahashi, H.: Nonradial Oscillations of Stars. University of Tokyo Press, Tokyo (1979)
Van Horn, H.M., Weidemann, V. (eds.): White Dwarfs and Variable Degenerate Stars. University of Rochester, Rochester (1979)
Vila, S. C.: Ph.D. Thesis, University of Rochester, Study of the Last Stages of Evolution of Solar Mass Stars (1965)
Warner, B. (ed.): Variable Stars and Galaxies. Astronomical Society of the Pacific, San Francisco (1992)
Watson Jr., T.J., Petre, P.: Father Son & Co. Bantam Books, New York (1990)
Wegner, G. (ed.): White Dwarfs. Proc. IAU Colloq, vol. 114. Springer, Berlin (1989)
Wegner, G.: Ph.D. Thesis, University of Washington, Seattle, The Spectra and Element Abundances of Cool White Dwarfs (1971)
Wesemael, F.: Ph.D. Thesis, University of Rochester, Atmospheres for Hot, High-Gravity, Pure Helium Stars (1979)
Winget, D.E., Van Horn, H.M.: *Sky & Tel.*, **64**, 216, "ZZ Ceti Stars: Variable White Dwarfs." (1982)
Winget, D.E.: Ph.D. Thesis, University of Rochester, Gravity Mode Instabilities in DA White Dwarfs (1981)
Wu, Y.: Ph.D. Thesis, California Institute of Technology, Excitation and Saturation of White Dwarf Pulsations (1998)

Index

A

AAS. *See* American Astronomical Society (AAS)
ABC. *See* Atanasoff-Berry Computer (ABC)
Aberdeen Proving Ground, 89, 93
Abrikosov, A., 109
Absolute bolometric magnitude (M_{bol}), 168, 169, 251
Absolute visual magnitude, 168–170, 220
Absorption (of radiation)
  bound-bound (line), 229
  bound-free, 27, 163, 229, 230
  coefficient, 153
  free-free, 27, 229
Abundances
  cosmic, 52
  of elements, 188
Accretion, 188, 189, 191, 193, 194, 241–245
  disk, 241–244
Acoustic wave, 198
Adams, Walter S., 15
Adiabatic
  displacement, 125
  expansion, 17
  gradient, 18, 125, 128
  pressure-density relation, 19
AGB. *See* Asymptotic giant branch (AGB) phase
Alias, diurnal, 212
α Canis Majoris B (Sirius B), 86
Alpha particle, 45, 50, 51
AM Canis Venaticorum, 204
American Astronomical Society (AAS), 157, 172, 182, 198, 215, 227, 232, 251
AM Herculis, 229, 244
Angel, J.R.P., 226, 228
Ångstrom unit, 12, 22, 34, 70, 77, 83, 84, 139–143, 148–150, 181
Angular momentum, 24, 33, 219, 220, 239, 246
Ap/Bp star, 233, 234
Apollo 11, 138
Apollo mission, 77, 144
Apollo-Soyuz, 144, 179
Apparent visual magnitude, 168, 169, 220
Archaeoastronomy, 176
Arequippa, Peru, 66
Armstrong, Neil, 138
Asteroseismology, 203, 211–223
Astrometry, 143
Astronaut, 138, 144, 146
Astronomical Netherlands Satellite (ANS), 179
Astronomical twilight, 216
Astronomical unit (AU), 3, 4, 237
Astrophysical Journal, 37
Astrophysics, 8, 25, 65, 82, 100, 101, 112, 116, 138, 159, 161, 175, 187, 268
Asymptotic giant branch (AGB) phase, 104–106, 183, 191, 235, 238
Atanasoff-Berry Computer (ABC), 88, 89
Atanasoff, John, 88–90, 94
Atkinson, Robert d'E., 42
ATLAS, 153–155, 157, 161
Atmosphere
  of Earth, 146
  solar, 25
  of star, 98, 153
  of Sun, 11
Atmospheric scintillation, 212
Atomic
  bomb, 57, 91–93, 96, 102
  mass, 44, 45, 265
  nucleus, 33, 42, 265
  number, 33
AT&T, 96
AU. *See* Astronomical unit (AU)
Auer, Laurence H., 153, 160, 180
Auer-Mihalas model atmosphere code, 161, 162
Australia, 50, 58, 72, 110

H.M. Van Horn, *Unlocking the Secrets of White Dwarf Stars*, Astronomers' Universe, DOI 10.1007/978-3-319-09369-7

B
Back warming, 154
Baglin, Annie, 186, 187, 198
Balmer, Johann Jacob, 22, 230
Balmer lines
  Hα, 77, 84
  Hβ, 84
  Hγ, 152
Bardeen, John, 72
Barnard, Dr. Frederick P., 5
Barstow, Martin, 147, 148, 180, 207
Baryon, 262, 268
Basel, 22
Basri, Gibor, 167
Beauchamp, Alain, 163, 167, 182, 206
Beckman Instruments, 213
Bell Telephone Laboratories, 261
Bergeron, Pierre, 162, 163, 166, 167, 176, 181, 182, 250, 251
Berry, Clifford, 88, 90
Bessel, Friedrich Wilhelm, 1–6
Bethe, Hans Albrecht, 42–45, 48, 50, 51, 54, 57, 92, 115
Biermann, Ludwig, 126
Big Bang, 256, 257, 261, 267, 268
BINAC, 96
Binary
  numbers, 88, 89
  pulsar, 32, 244
  star (evolution), 6, 7, 17, 75, 105, 235, 237, 238, 244, 254
Blink comparator, 65
Blue horizontal branch, 254
Body-centered cubic (bcc) lattice, 109, 112
Böhm, Karl-Heinz, 126–129, 157, 177, 184
Böhm-Vitense, Erkia, 126–128, 158
Bohr, Niels, 24
Bok, Bart, 184
Bolometric correction, 168, 169, 251
Boltzmann, Ludwig, 12
Bond, Howard, 147
Boyle, Robert, 18
BPM 25114, 231
Bradley, James, 2
Branch, David, 264
Brattain, Walter, 72
Bremen, 1, 2
Brickhill, A.J., 208, 209
Brownian motion, 29
Bruce proper motion (BPM) survey, 66
Brush, Stephen G., 111, 112
Bubble, 125, 126
Bues, Irmila, 157
Buoyancy, 198
Burbidge, E. Margaret, 100
Burbidge, Geoffrey, R., 100

C
Calar Alto, 72
California, 63, 75, 76, 78, 79, 83, 99, 102, 111, 165, 174, 175
California Institute of Technology (Caltech), 79
Cambridge, England, 100, 157, 175, 184
Cambridge, Mass, 9
Cannon, Annie Jump, 13
Capacitor, 88, 90
Carbon-oxygen core (C/O core), 104, 107, 143, 194, 238
Carnegie, Andrew, 79
Carnegie Institution of Washington, 44, 78, 181, 262
Cataclysmic variable (helium), 229, 239–241, 244, 245, 254
Cavendish Laboratory, 43
Cayuga Lake, 114
CDC. *See* Control Data Corporation (CDC)
CDC 6600, 98
Celestial coordinates, 3
Central star of planetary nebula (CSPN), 105, 106, 115, 184, 187, 192, 193, 207, 218
Centre de Recherche en Astrophysique du Québec (CRAQ), 162
Centre National de la Recherche Scientifique (CNRS), 123, 188
Cepheid
  instability strip, 199
  variable, 199, 208, 259, 263, 264
Cerro Tololo Interamerican Observatory (CTIO), 71, 196, 266
Challenger (STS-7), 143
Chandrasekhar limit, 37, 39, 103, 110, 245, 246, 264, 266
Chandrasekhar, Subramanyan, 19–27, 35–39, 41
Chanmugam, Ganesh, 159, 197
Charge-coupled device (CCD) array, 72, 73, 148, 165, 167, 169, 254
Chemical
  composition, 11, 101, 117, 151, 177–181, 263
  constituent, 123
  enrichment, 121
  picture, 132
  segregation, 120
Circular polarization, 226

Circumstellar gas, 143
City College of New York (CCNY), 54
Civil War, American, 5
Clark, Alvan, Jr., 5, 6, 63, 76
Clark, Alvan, Jr. & Sons, 63
Close-binary
evolution, 239, 240
orbit, 234
system, 238–240, 244, 245, 264
CMB. *See* Cosmic microwave background (CMB)
CMD. *See* Color-magnitude diagram (CMD)
CNO
bi-cycle, 46, 48
elements, 121, 193
CNRS. *See* Centre National de la Recherche Scientifique (CNRS)
COBE. *See* Cosmic Background Explorer (COBE)
Cold War, 57, 115
Color-magnitude diagram (CMD), 170, 253, 255
Colors of stars, 151
Comet, Halley's, 191
Comet Shoemaker-Levy 9, 190
Common envelope, 239, 240, 243, 246
Composition transition zone, 189, 194, 218
Computer
algorithm, 99, 196
code, 117, 153
digital, 49, 88, 111, 114, 153, 198, 213
electronic digital, 88, 90, 91, 94, 98, 99, 102
laptop, 216
subroutine, 117
technology, 212
terminal, 140
Computing Tabulating Recording (CTR) Corp., 92
Conductive opacity, 55, 102, 115
Conductivity
electrical, 53, 69
thermal, 54, 55
Conservation of mass, 18, 27
Constellation
Orion, 13, 189, 252
Sagittarius, 247
Continuum emission, 154, 225
Control Data Corporation (CDC), 67, 97, 98
Convection
turbulent, 125
zone, 123, 125, 128, 129, 132–135, 177, 184, 192–194, 208, 209, 235
Convective
coupling, 133, 252
element, 126, 128
energy transport, 126–128
envelope, 85, 105, 235
flux, 157
mixing, 143, 186, 192, 206, 209
Cooke refractor, 68
Cool, Adrienne M., 256
Copernicus (satellite), 139
Corning Glass Works, 79
CoRoT spacecraft, 188
Corrective Optics Space Telescope Axial Replacement (COSTAR), 146
Cosmic Background Explorer (COBE), 262
Cosmic background radiation, 261
Cosmic microwave background (CMB), 261–263, 267
Cosmological
constant, 260, 266–268
distance, 256
model, 261, 267
redshift (z), 263
Cosmology, standard model of, 179, 259–269
Cosmonaut, 138, 144
Coulomb interaction, 109, 129, 131
Cousins, Alan, 70
Cowling, Thomas G., 198, 208, 233
Cox, Arthur N., 102, 132
Crab Nebula, 254, 255
Crafoord Prize, 52
CRAQ. *See* Centre de Recherche en Astrophysique du Québec (CRAQ)
Cray, Seymour, 97, 98
Critchfield, Charles L., 44
Crystalline
lattice, 111, 113
solid, 111, 113
Crystallization, 111–114, 117–120, 252, 257
Crystallization (curves), 117
CSPN. *See* Central star of planetary nebula (CSPN)
CTIO. *See* Cerro Tololo Interamerican Observatory (CTIO)
Cyclotron orbit, 230

D
Dahn, Conard C., 170, 250
Damping mechanism, 203
D'Antona, Francesca, 133
DAO white dwarf, 207
Dark energy, 266–268
Dark matter, 262, 266–268
Dartmouth College, 178
Data processing, 73
Daughter mode, 209
DAV white dwarf, 223
DA white dwarf, 83, 133, 143, 144, 149, 152–156, 158, 162, 166, 167, 171, 176, 179–184, 186, 191–193, 200, 203, 205
DAZ white dwarf, 192, 193
DBAP3 white dwarf, 227
DBA white dwarf, 167, 194
DB gap
  blue edge of, 193
  red edge of, 192
DB white dwarf, pulsating, 83, 163, 182, 184, 192, 194, 204–206
DC white dwarf, 143, 186, 193
de Broglie, Louis, 32
Debye
  cooling, 111, 113, 114, 117, 119
  temperature, 12, 111, 113, 118, 119
Deceleration parameter, 261
Declination, 3, 68, 71, 86
Defense Calculator, 96
Degeneracy boundary, 56, 60, 129, 133, 134
Degenerate
  core, 58–60, 111, 129, 133, 134, 221, 252
  electrons, 37, 53, 55, 57, 58, 60, 115, 131, 265
  matter, 38, 53, 60, 265
  model, 53
  star (variable), 39, 54, 57, 58, 71, 180, 183, 186, 192, 199, 206, 207, 211, 220, 223, 231, 265
Delaware Asteroseismology Research Center, 173
Demarque, Pierre, 101
Density, 11, 16–20, 27, 34, 37, 38, 44, 55, 57, 101, 111, 112, 125, 131, 132, 153, 208, 243, 248, 264–266, 268
Department of Terrestrial Magnetism (DTM), 44, 264
Deuterium, 44, 150, 267
Deuteron, 44
DF white dwarf, 84
DG white dwarf, 84
Diamond, 109–121
Diffusion (coefficients), 55, 188, 189, 192–194
Digital computer
  electronic, 88, 90, 91, 94, 98, 99, 102
  programmable, 87
Diode, 90, 97
Dirac, Paul A.M., 35
Discovery (STS-31), 145
Dolez, Noel, 202, 209
Donor star, 242–245
Doppler shift, 174
Double white dwarf binary, 244, 245
DOV star, 206, 217
DO white dwarf, 144, 193
DQ Herculis, 229, 245
DQ white dwarf, 83, 194
Draper, Henry, 12–14
Drunkard's walk, 25
Dry ice, 70
Dwarf nova, 240–244
Dziembowski, Wojtek, 200, 202
DZ white dwarf, 83

E
Eagle Nebula, 189, 190
Earth, 1, 3, 6, 9–17, 20, 33, 41, 52, 128, 137, 138, 140, 143, 144, 146–149, 173, 174, 188, 211–223, 225, 247, 265
Eastman Quadrangle, 112
EC 20058+5234, 223
Eckert, J. Presper, 90, 94
Eddington, Arthur Stanley, 16, 26
Edinburgh-Cape (EC) Blue Object survey, 71
Effective temperature (of a star), 47–49, 70, 83, 84, 105, 106, 128, 133, 140, 143, 144, 151, 152, 155, 161, 165–168, 171, 172, 180, 181, 184, 186, 191–194, 199, 203, 204, 206, 207, 209, 218, 220, 227, 252
EG049 (Sirius B), 86
Eggbeater, 239
Eggen, Olin J., 82, 86
Egyptology, 176
Einstein, Albert, 24, 29, 30, 39, 80, 260
Einstein Observatory (HEAO-2), 143
Eisenhower, President Dwight D., 138

Electron, 24–27, 32–38, 44, 50, 53–55, 57, 58, 60, 72, 91, 109, 110, 115, 117, 124, 129, 131–134, 155, 163, 194, 230, 265
  conduction, 55, 133, 134, 194
Electronic Discrete Variable Automatic Computer (EDVAC), 94
Electronic logic circuits, 88
Electronic Numerical Integrator and Computer (ENIAC), 90, 91, 94–96
Electronics, 50, 69, 72, 73, 80, 88–91, 94–99, 102, 196, 212
Electrostatic interaction, 35, 109, 110, 112
Elementary particle physics, 57
Elements, chemical
  calcium (Ca), 11, 84, 183, 188
  carbon (C), 45, 84, 85, 143, 188, 192
  deuterium (D =^{2}H), 44
  helium (He), 11, 13, 45, 132, 179
  hydrogen (H), 11, 44, 45, 132, 179
  iron (Fe), 11, 84, 143, 188
  lithium (Li), 267
  magnesium (Mg), 11, 188
  neon (Ne), 240
  nickel (Ni), 4
  nitrogen (N), 45, 180, 192
  oxygen (O), 45, 143, 180, 192
  phosphorus (P), 227
  potassium (K), 70
  silicon (Si), 143, 192
  sodium (Na), 11
  sulfur (S), 180
Elkin, W.L., 4
E-mail, 215
Emergent intensity, 230
Endeavour (STS-61), 146
Energy, 11, 19, 23–28, 30–33, 35, 37, 38, 41–44, 46, 48, 50–52, 55, 57–60, 87, 88, 100, 101, 103, 104, 109, 111–113, 115, 116, 119, 120, 124–126, 128, 129, 131, 133, 134, 153, 155, 157, 179, 196, 207–209, 226, 227, 229–231, 237–243, 245, 263, 265–268
Energy-generation rate, 58, 265
ENIAC. *See* Electronic Numerical Integrator and Computer (ENIAC)
Entropy, 101
Envelope (of star), 27, 218
  model, 129, 133, 135
Equation of state, 19, 35, 38, 39, 101, 102, 109, 110, 115, 117, 125, 129, 131–133, 162, 229
Equilibrium
  convective, 18, 25
  radiative, 25, 26
ESO. *See* European Southern Observatory (ESO)
European Southern Observatory (ESO), 72
EUVE. *See* Extreme Ultraviolet Explorer (EUVE)
EUVE J0317-853, 234
Evolution
  of star, 99, 101, 105, 106, 176, 235
  stellar, 22, 37, 99–106, 114, 117, 118, 188, 237, 239, 240, 253
Excitation mechanism, gamma
  convective driving, 208
  kappa, 208
Exclusion principle, 33, 35
EXOSAT, 143, 207
Explorer I, 137
Exposure, 72, 80, 147, 148, 195, 212, 255
Extreme ultraviolet (EUV), 144, 149, 179, 180
Extreme Ultraviolet Explorer (EUVE), 148, 149, 167, 234
Fahlman, Gregory G., 254
Faint-object photometry, 149
Faint object spectrograph (FOS), 146
Fairchild Semiconductor, 72
Far ultraviolet (FUV), 150, 179
Far Ultraviolet Spectroscopic Explorer (FUSE), 139, 149, 150
Fast Fourier Transform (FFT), 196, 211
Faulkner, John, 243
Feige 4, 83, 84
Feige 7, 227
Feige 24, 144
Fermi
  energy, 35, 55, 57, 60, 265
  level, 37
  momentum, 35, 38
Fermi-Dirac statistics, 35, 36
Fermi, Enrico, 34, 35
Ferrario, Lilia, 228, 232–235
Feynman, Richard P., 91–93
Field-spread broadening (of absorption line), 230
Filaments (in cosmology), 262
Fine Guidance System (FGS), 148
Finger Lakes, 114

Fingers of God, 189, 190
Finley, David, 167
Fireball, 102
First light, 64
Fleming, Mrs. Williamina, 14, 15
Florida Institute of Technology, 250
Fluid, 17, 18, 109, 112, 113, 119, 120
Folk song, 131
Fontaine, Francine, 130
Fontaine, Gilles, 130, 132–134, 160–162, 180, 188, 189
Ford, W. Kent, Jr., 262
FORTRAN, 212
40 Eridani A, 14
40 Eridani B, 14, 15, 23, 64, 152, 155, 156, 172, 176, 246
47 Tucanae (star cluster), 149
Fowler, Ralph H., 34–37, 39
Fowler, William A., 52, 100
Frankel, Stanley, 92
Fraunhofer, Joseph, 9–11
Free energy, 131
Fulbright, Mike, 176

G
Gagarin, Yuri, 138
Galactic
  bulge, 247, 248
  center, 248
  disk, 247–255, 272
  halo, 254
  protodisk, 257
Galaxy(ies), 121, 66, 73, 103. 111, 149, 176, 179, 190, 245–257, 260–264, 267
γ4670 star, 85
Gamma (γ) ray, 45, 100, 208
Gamow, George, 42, 44, 45, 95, 261
Garstang, Roy, 226, 227
Gay-Lussac, Joseph Louis, 18
GD 52, 83, 84
GD 154, 223
GD 358, 204–206, 222
GD stars, 68
General Electric, 96
General relativistic theory of gravitation, 174, 175, 244, 260, 266, 267
General relativity, 32, 261
Germany, 2, 29, 43, 50, 101, 151, 185, 187, 250
Giant planets, 120
Giclas, Henry, 68
Gigayear (Gyr), 251, 252
Globular star cluster, 149, 248, 253–257
G-numbered stars, 68
Goddard high resolution spectrograph (GHRS), 146
Goldreich, Peter, 209
Goldstine, Herman, 89, 90, 93, 94
Goudsmit, Samuel, 33, 43
Graboske, Harold C. Jr., 131, 132
Gradient, temperature, 17, 18, 25, 53, 55, 120, 124, 125, 128, 133, 157, 177
Graham, John, 181
Gravitation(al)
  potential energy, 19, 120, 238
  radiation, 32, 244, 246
  redshift, 31, 83, 140, 143, 147, 148, 150, 174–176
  segregation (of elements), 180, 189, 193, 194
  waves, 32, 175, 244
Great Attractor, 179
Great Depression, 82
Great Nebula in Orion, 254
Great Observatories, 145
Great Refractor, 64, 76
Green, Richard, 71, 86, 206
Greenstein, Jesse L., 80–84, 86, 140, 152, 165, 168, 174, 176, 183, 186, 188, 225, 228
Greenwich Mean Time, 147
Grenfell, Thomas C., 157, 158, 162
Grw+70° 8247, 225, 226, 228, 229
GW Virginis, 207

H
H1504+65, 143
Hale, George Ellery, 76, 78–80
Hale Telescope, 79, 81
Hamada, T., 110
Hamburg-ESO (HE) survey, 72
Hamburg-Schmidt (H-S) survey, 72
Hanford, 212
Hansen, Brad M.S., 255, 256
Hansen, Carl J., 198–200, 208, 241
Harmonice Mundi, 7
Harmon-Seaton sequence, 105
Härm, Richard, 101
Haro, Guillermo, 69, 196
Haro-Luyten Taurus 76, 196
Harper, William Rainey, 76, 78
Harris, Hugh C., 252
Harvard Annals, 14
HD149499B, 144

Heat content, 17, 59, 60, 101, 111, 113, 157
Heisenberg, Werner Karl, 34
Helium
  burning, 49, 50, 58, 103–105, 183, 235, 237–240, 251
  rich atmosphere, 143
Helmholtz, Hermann von, 41
Henry Draper Catalogue, 13, 14
Henry Draper Memorial Fund, 12
Henyey, Louis G., 99
  method, 99–101
Herschel, Sir John, 4
Hertzsprung, Ejnar, 15, 20, 22, 63, 75
Hertzsprung-Russell diagram (H-R diagram), 22, 23, 47–49, 103, 104, 169, 188, 199, 254
Hesser, James E., 196, 197
High-field (magnetic "spot"), 231
High-redshift, 266
High-speed
  photometer, 214
  photometry, 204, 206, 213, 214, 216, 217
High-z Supernova Search Team, 264, 267
Hipparcos (satellite), 143, 169
History of science, 161, 176
Holberg, Jay, 4, 5, 7, 8, 11, 14, 18, 21, 22, 27, 31, 33, 34, 36–38, 46, 54, 67, 141, 144, 147–150, 167, 237
Hollerith card, 92
Hot spot, 241, 242
Hot subdwarf
  sdB, 246
  sdO, 246
Houtermans, F.G., 42
Hoyle, Sir Fred, 52, 100
HST. *See* Hubble Space Telescope (HST)
Hubble
  constant (H0), 261, 268
  diagram, 266
Hubble, Edwin P., 80, 260, 261
Hubble Space Telescope (HST), 145–147, 149, 167, 171, 189, 254–256
Hulse, Russell, 244
Humason, Milton, 68, 69
Hyades, 68, 69, 254
Hydrides, 70
Hydrogen
  bomb, 102
  burning, 46–48, 57, 58, 103, 104, 106, 183, 237, 239, 240
  lines, 11, 14, 15, 22, 58, 83, 166, 167, 184, 228
Hydrogen-lines (H-lines). *See* Balmer lines
Hydrostatic equilibrium, 17, 18, 26, 27, 87, 125, 161
HZ 43, 179, 180

I

IAU Colloquium 56, 199–200
Iben, Icko, Jr., 100, 105, 109, 183, 201, 239, 246
Iben, Icko, Sr., 100
Ice, 70, 113, 219
Identification charts, 69
Iglesias, Carlos A., 102
Implosion, 92, 93
Indian Ocean, 148
Infrared Telescope Facility (IRTF), 251
Inner Lagrange point, 238, 241
Instability strip, 199, 200, 204, 206, 209
  Cepheid, 199
  DBV
    blue edge, 206
    red edge, 206
  ZZ Ceti
    blue edge, 204
    red edge, 209
Institut d'Astrophysique, 187
Institute for Advanced Study (Princeton), 95
Institute for Astronomy (Cambridge), 157
Institute for Theoretical Physics (Kiel), 151
Integrated circuit, 72, 97
International Business Machines (IBM), 92–98
  cards, 198
  IBM 650, 98
  IBM 701, 96
  IBM System/360, 97, 98
International Ultraviolet Explorer (IUE), 139, 141–143, 167
Interstellar
  cloud, 190, 191, 252
  grains, 82
  matter, 189, 191, 193, 194
Ion, 25, 117, 153, 226
Ionization
  equilibrium, 129, 131
  potential, lowering of, 155
  pressure, 27, 34, 131, 155
  thermal, 35
Iowa State College, 88, 89

Iron peak elements, 265
Isern, Jordi, 121
Ithaca, New York, 43, 44, 114
IUE. *See* International Ultraviolet Explorer (IUE)

J
Jewish law, 176
JOHNNIAC, 96
Johnson, Harold L., 70
Joint Institute for Laboratory Astrophysics (JILA), 198, 226
Jupiter (planet), 190
Jupiter C, 137

K
Kawaler, Steven D., 199, 201, 215, 217, 218, 220, 221, 241
Kellogg Radiation Laboratory, 52, 100
Kelvin, Lord. *See* Thomson, William
Kemic, S.B., 226–228
Kemp, James, 225
Kennedy, President John F., 138
Kepler, Johannes, 7
Kiel model atmosphere code, 161
Kilby, Jack, 72
Kilogauss (kG), 225
Kiloparsec (kpc), 247, 248, 254–256, 262
Kippenhahn, Rudolf, 101, 238
Kirchoff, Gustav (laws of radiation), 11, 12
Kirshner, Robert P., 264
Kirzhnits, D.A., 109
Koester, Detlev, 105, 158, 159, 161, 167, 176, 181, 182, 202, 250
Königsberg, 2, 3, 5, 41
Kron, Gerald, 70
Kuiper, Gerard P., 75–77, 184
Kurucz, Russell, 153, 155

L
L19-2, 223
L879-14, 83, 85
Laboratoire d'Etude Spatiale et d'Instrumentation en Astrophysique (LESIA), 188
Lambda Cold Dark Matter (ΛCDM), 267
Lamb, Donald Quincy, Jr., 115–117
Landau
  continuum, 230
  level, 230
Landolt, Arlo, 159, 195, 196
Landstreet, John, 135, 226
Lane, J. Homer, 18, 19
Laser Interferometer Gravitational-wave Observatory (LIGO), 244
Lasker, Barry M., 196, 197
Latent heat, 113, 114, 118–120, 257
Lawrence Berkeley Laboratory (LBL), 264
Lawrence Livermore National Laboratory (LLNL), 102, 111, 131
LB 1497, 103
Leavitt, Henrietta, 259
Lee, T.D., 58
Leggett, Sandy, 251, 252
LFT0486 (Sirius B), 86
Lick, James, 63
Liebert, James W., 71, 165–167, 176, 186, 207, 227
Light-collecting "buckets", 72
Light, speed of, 31, 35, 91
Lillienthal, 2
Line blocking, 154
Line-formation theory, 178
Line profile, 152, 154, 166, 178, 182
Line-shift broadening (of absorption line), 230
Local dawn, 214, 215
Local Group (of galaxies), 179
Local thermodynamic equilibrium (LTE) model, 162
Longitude, 197, 201, 214, 217
Los Alamos National Laboratory (LANL), 57, 102, 132, 245
Luminosity, 14, 15, 20, 58, 60, 61, 87, 99, 104, 105, 118, 119, 128, 129, 133, 134, 150, 166, 168, 171, 192, 195, 196, 208, 220, 222, 234, 240, 241, 243, 249, 250, 252, 255, 259, 263, 264, 266
  of Sun, 41, 46–49, 53, 119, 135, 218
Lunar
  farside, 138
  surface, 138
Lunar and Planetary Laboratory (LPL), 77
Lunar Excursion Module (LEM), 138
Luyten Blue (LB) catalog, 69
Luyten Half Second (LHS) catalog, 66
Luyten Palomar (LP) survey, 66, 67

Luyten Two Tenths (LTT) survey, 66
Luyten, Willem J., 63–69, 75, 77, 78, 86, 155–156, 166, 169, 196
Lyman alpha (Lα) lines, 140

M

M1 (Crab Nebula), 254
M4 (star cluster), 149
M31 (Andromeda Galaxy), 254
M42 (in Orion), 254
M43 (in Orion), 254
M71 (star cluster), 254
Magneto-bremsstrahlung, 229
Magni, G., 132
Magnitude
  absolute, 15, 20, 23, 47, 61, 151, 170, 186, 249–250, 254, 255
  apparent, 3, 14, 20, 64–66, 70, 155, 220, 255, 266
Main Sequence, 21–23, 46–48, 51, 52, 64, 68, 71, 75, 77, 82–84, 103–106, 149, 153–155, 183, 199, 219, 220, 231, 233–240, 246, 254, 256
Manhattan project, 57, 102
MANIAC, 95
Manned space program, 144
Marchant, 92
Mark, J. Carson, 95
Marshak, Robert E., 53–57, 115, 116
Martin, B., 207, 230–232
Massachusetts Institute of Technology (MIT), 76, 78, 101
Mass-energy, 42, 46, 266, 268
Mass loss, radiation-driven, 104–106, 184, 192, 193, 235
Mass-radius relation (for white dwarfs), 39, 53, 109, 110, 152, 155, 171–173, 181
Mather, John, 263
Matsushima, Satoshi, 154, 155
Mauchly, John, 90, 94, 95
Mazzitelli, Italo, 132, 133
McGraw, John T., 197, 206, 207, 214
MCT 0455-2812, 180
Megagauss (MG), 225, 226, 235
Megaparsec (Mpc), 263
Merger scenario (for origin of magnetic white dwarfs), 234
Meridian circle, 3
Meson physics, 57
Messier, Charles, 254
Mestel, Leon, 58–6, 110, 111, 113–115, 119
Metallization (of hydrogen), 131
Metals, 24, 34, 54, 85, 94, 131, 178, 188
Metropolis, Nicholas, 95
Michaud, Georges, 189
Microwave, 261, 262, 267
Mihalas, Dimitri, 153, 162
Milky Way Galaxy, 66, 111, 179, 190, 245, 247–257
Milne, E.A., 37
MIT. *See* Massachusetts Institute of Technology (MIT)
Mixing length theory, 125, 126, 128, 158, 206, 208
Mochkovitch, Robert, 120, 121
Model atmosphere, 129, 140, 153–163, 167, 168, 177, 179–181, 206, 207, 230, 251
Mode splitting, 219
Molecular C_2, 85, 143
Monet, David G., 250
Monochromatic intensity, 87
Monthly Notices of the Royal Astronomical Society, 4
Montreal-Cambridge-Tololo (MCT) survey, 71
Moon, 77, 138, 144
Moore School of Engineering, 90
Morgan, William W., 70
Multi-mode pulsator, 203
Multiplet, 222

N

Narrow-band colors, 163
NASA. *See* National Aeronautics and Space Administration (NASA)
Nather, R. Edward, 197, 204, 212–216
National Advisory Committee for Aeronautics (NACA), 138
National Aeronautics and Space Administration (NASA), 77, 138, 139, 142, 145, 149, 190, 242, 251
National Radio Astronomy Observatory (NRAO), 82
National Research Council, 82
National Science Foundation (NSF), 67, 107, 248, 253
Naval Ordnance Laboratory, 94
Nazi (Germany), 43, 50, 123
Near infrared, 70, 146
Near ultraviolet, 146
Netherlands, 63, 75, 112
Neutral atomic carbon (C I), 85, 143

Neutrino (energy-loss rate), 44, 45, 115, 117, 221
Neutron star, 16, 234, 246
New General Catalog (NGC), 105, 149, 255–257
New Luyten Two Tenths (NLTT) catalog, 66
Newton, Sir Isaac, 7, 9
New York City, 12, 80, 169
NGC 6397 (star cluster), 149, 256, 257
NGC 6752 (star cluster), 149, 254, 255
Nobel Prize, 38, 45, 91, 244, 259–261, 263
Non-DA white dwarf, 133, 171, 183, 186, 191
Non-degenerate envelope, 58, 60, 218
Non-ideal effects, 131
Nonlinear energy transfer, 209
Norris, William, 97
Nova
  classical, 241, 245
  dwarf, 240–244
Noyce, Robert, 72
NRAO. *See* National Radio Astronomy Observatory (NRAO)
NSF. *See* National Science Foundation (NSF)
Nuclear
  energy, 28, 42, 43, 46, 48, 104, 196, 237, 265
  reaction, 42, 44–46, 48, 58, 101, 103, 104, 112, 114, 115, 212, 251, 264, 265, 267
  reaction rate, 101, 251
  reactor, 212
Nucleogenesis, 101
Nucleus, 24, 33, 42, 44–46, 50–52, 109, 114, 265

O

OAO. *See* Orbiting Astronomical Observatory (OAO)
Objective
  lens, 5
  prism, 13, 72
Observatory
  Amtmann Schroeter's, 2
  Apache point, 73
  Cerro Tololo Interamerican (CTIO), 71, 196, 197, 266
  David Dunlap, 101
  Harvard College, 9, 10, 12, 13, 21, 65, 66, 69, 75, 259
  Kitt Peak National (KPNO), 177, 195
  Lick, 63–65, 75, 76, 206, 227
  Lowell, 68, 69
  McDonald, 197, 204, 205
  mountaintop, 63, 205
  Mt. Palomar, 78, 140
  Mt. Wilson, 78, 100
  Nice, 123, 187
  Pulkovo, 7
  Steward, 184, 206, 227
  Sutherland, 197
  Tonanzintla, 69
  University of Minnesota, 66
  U.S. Naval (USNO), 169, 170, 250, 252
  U.S. Naval, Flagstaff Station, 250
  Yale University, 4, 153
  Yerkes, 39, 64, 75, 78, 79, 82, 153
OCP. *See* One-component plasma (OCP)
Ohmic dissipation, 233
Oke, J. Beverley, 155
Ω Centauri (star cluster), 149
One-component plasma (OCP), 111, 112
Opacity
  bound-free, 230
  free-free, 162
Opacity project (OP), 102, 103
OPAL, 102, 103, 132
Oppenheimer, J. Robert, 96
Optically thin (atmosphere), 128, 157
Orbital period, 5, 7, 240, 244, 245
Orbiting Astronomical Observatory (OAO), 139
Orbiting Solar Observatory (OSO), 138
Orion
  constellation, 6, 13, 189, 252, 254
  Nebula, 189
Osaki, Yoji, 198, 199, 242
Oscillation
  g-mode, 198, 217
  high-order, 217
  mode, 198, 201–203, 205, 211, 212, 216–219
  multiperiodic, 207
  p-mode, 198
OSO. *See* Orbiting Solar Observatory (OSO)
Overtone, 198, 217

P

Paczynski, Bohdan, 101
Palomar Observatory Sky Survey (POSS1), 67

Paquette, C., 189
Parallax, 3, 4, 14, 15, 17, 21, 144, 155, 168–170, 249
Parsec, 4, 222
Partial ionization zone, 124, 132, 199, 203, 204, 208
Particle accelerators, 42
Pasadena, 79, 80
Pauli, Wolfgang, 33, 35
Peierls, Sir Rudolf, 50
Penzias, Arno A., 261
Periodicity, 202, 211, 212, 221
Period, pulsation, 195–200, 208, 221, 259
Perlmutter, Saul, 259, 260, 263, 264, 266, 267
Peters, Christian August Friedrich, 5
PG 1031+234, 226, 231, 232, 235
PG 1159-035, 71, 143, 206, 207, 215, 217, 218, 220, 221
PG 2131+066, 220, 223
PG 1159 star, 84, 193, 217, 218, 221–223
Phase
- diagram, 119, 120
- separation, 118–121, 252
- transition, 112, 113, 117, 257
  - first-order, 113, 257
  - fluid/solid, 112

Photoconductive device, 69
Photoelectric
- cells, 70
- effect, 24, 29, 32, 70
- magnitude, 70
- photometry, 70

Photographic plate, 12, 13, 15, 65–68, 71–72
Photography, 1
Photometer
- three-channel, 215, 216
- two-star, 216

Photometry/photometric, 69–71, 149, 151, 168, 170, 181, 182, 195, 197, 200, 204, 206, 211–217, 254, 255
Photomultiplier tube, 70
Photon, 24–27, 29, 70, 72, 188, 212
- diffusion, 55

Photosphere, 141, 192–194, 230
Physical picture, 132
Pickering, Edward C., 9, 10, 12–15, 21, 259
Pillars of creation, 189
Pixel, 72, 73
Planck, Max, 22–24
Planck's constant, 24, 32, 34
Planetary nebula, 106, 107, 183, 188, 207, 235, 238
Planetary nebula nucleus variable (PNNV) star, 207, 223
Plasma
- one-component (OCP), 111, 112
- partially ionized, 131

Pleiades, 20, 103
Pluto, 68
Polar (intermediate), 243, 245
Polytropic model, 19, 25–27
Positron, 44, 45
Post-AGB star, 105, 106, 183, 184, 192, 218
Potential barrier, 42
Power spectrum, 211, 212, 222
Prandtl, Ludwig, 125, 126
Pressure (ionization), 17–19, 26, 27, 34–38, 58, 60, 87, 99, 101, 125, 126, 155, 161, 174, 188, 198, 208, 230, 241, 265
Prism, 9, 12, 13, 72, 75
Probability (of nuclear reaction), 42, 52
Procyon, 4, 5
Procyon B, 171, 172
Proper motion, 5, 14, 64–69, 77, 85, 86, 143, 250
Proton, 33, 44–46, 48
Proton-proton chain, 44, 46, 48
Protostar, 253
Provencal, Judi, 171–173, 176, 216, 222
Ptolemy, 1
Puebla, Mexico, 69
Pulsation
- mode, 196, 200, 202, 217, 221, 223
- period, 195, 198–200, 208, 221, 259
- radial, 196

Pycnonuclear reaction, 112, 265
Pyrex, 79

Q

Quanta, 23, 24
Quantum
- energy level, 155
- many-body problem, 132
- mechanics, 34, 42
- numbers, 33, 226, 227
- physics, 16, 32–34
- statistics, 28, 38
- theory, 22, 25, 31, 32

Quasar (QSO), 71–73, 86
Quiescence, 240, 243

R
R 548, 197, 214
RA. *See* Right ascension (RA)
Radial
  overtone, 217
  velocity, 143, 176, 260
Radiant energy, 27, 101
Radiation
  belts, 137
  black-body, 11, 12, 22, 262
  pressure, 160, 161, 188
  thermal, 11, 12, 24, 261, 262
Radiative
  levitation, 192
  transfer, 55, 87, 128, 157, 229, 230, 242, 243
Radio astronomy, 82, 244
Radiogenic age, 46
Rayleigh scattering, 163
RCA, 70, 96
Red dwarf, 14
Red giant branch (RGB), 49, 52, 103, 104, 106, 238, 239, 254
Redshift, 176, 262, 263, 266–268
  gravitational, 32, 83, 140, 143, 147, 148, 150, 174–176
Refractor (refracting telescope), 5, 63, 68, 76
Reid, I. Neill, 176
Reionization, 268
Relativity
  general theory, 31, 83
  special theory, 30, 31
Remington Rand, 95, 97
Renzini, Alvio, 105, 106, 254–256
Resistor, 90, 97
Rest mass, 46, 50
Reticle, 3
RGB. *See* Red giant branch (RGB)
Rhodes, Richard, 93–95
Richer, Harvey B., 237, 254, 255
Ride, Sally, 145
Riess, Adam, 259, 260, 264, 267
Right ascension (RA), 3, 66, 71, 86
Ritter, August, 19
Robinson, Edward L., 197–200, 204, 213, 214
Roche
  lobe, 238–242, 244, 246
  model, 237
Roche, Èdouard A., 238
Rochester conference, 57, 115
Rockefeller Foundation, 79
Rocket, 97, 137, 138, 179
Roentgen satellite (ROSAT), 149, 180
Rogers, Forrest J., 102, 103, 132
Roman, Nancy Grace, 138
ROSAT. *See* Roentgen satellite (ROSAT)
Rotation
  period, 197, 205, 212, 219, 220, 223, 229, 234, 235
  rate, 143, 219
  synchronous, 245
Rotational splitting, 219, 222
Royal Astronomical Society, 4, 37, 255
Royal Swedish Academy of Sciences, 52
Rubin, Vera, 262
Ruderman, Malvin A., 111, 113, 114
Ruiz, María Teresa, 251
Russell, Henry Norris, 14, 21–23
Russia, 7, 54
Rutherford, Ernest, 24, 42

S
Saffer, Rex, 166
Sahlin, Harry L., 111
Salpeter, Edwin E., 38, 50–52
Sandage, Alan, 261
Sarnoff, Gen. David, 96
Satellite (data), 137–140, 143, 149, 167, 169, 179
Sauvenier-Goffin, E., 196
Savedoff, Malcolm P., 112–114, 140, 160
Schatzman, Evry, 123, 124, 186, 189, 196, 198
Schmidt, Bernhard, 67
Schmidt, Brian P., 259
Schmidt, Gary, 231
Schmidt, Maarten, 71
Schmidt telescope
  Curtis, 71
  UK, 72
Schrödinger, Erwin, 32, 42
Schulz, H., 158, 181, 182
Schwarzschild, Karl, 25, 124
Schwarzschild, Martin, 101, 114
Science fiction, 137
SDSS. *See* Sloan Digital Sky Survey (SDSS)
Seaton, Michael J., 102, 105
Seismic distance, 220, 222
Semi-major axis, 7
Seven Samurai, 179
Sforzando event, 222
Shell
  flashes, 104, 106, 184
  source, 48, 104, 105, 183, 235
Shipman, Harry, 156, 157, 161, 173, 179, 180, 184, 191, 250

Shockley, William, 72, 96
Shot noise, 212
Silicon
chip, 72
wafer, 72
Sine curve, 195
Sion, Edward M., 85, 186
Sirius B, 4, 6–9, 15, 16, 34, 55, 56, 61, 63, 64, 86, 139–141, 144, 147–149, 160, 167, 171, 172, 179, 237
Sirius/Sirius A, 4, 6, 7, 15, 16, 140, 141, 144, 150, 179, 237
61 Cygni, 4
Skylab, 144
Sloan Digital Sky Survey (SDSS), 73, 243, 252
Smiley, Jane, 88, 90, 92, 94, 95
Smith College, 173, 174
Smith, George, 72
Smoot, George, 263
Snow, Ted, 139
Soap opera, 131
Soft X-ray, 143, 149, 179, 180
Solar System, 17, 24, 46, 121, 138, 190, 248, 265
Solid-state
devices, 72, 97
physics, 54, 55, 112, 113
Sommerfeld, Arnold, 36, 43
Soviet Union, 109, 115, 137
Space
age, 137
shuttle, 145
Spacecraft, robotic, 138
Space Telescope Imaging Spectrometer (STIS), 147, 148
Space Telescope Science Institute (STScI), 147, 148
Space-time, 31
Space Transportation System (STS), 145
Sparks, Warren, 245
Spectra
low-resolution, 147, 148
metallic line, 143
Spectral
classes of stars, 13, 21, 22, 85
classification, 13, 14, 22, 65, 71, 77, 78, 177
lines, 24, 25, 33, 177, 220, 226, 228, 263
Spectrograph, 12, 82, 140, 141, 146, 147
Spectroheliograph, 78
Spectrophotometry, 146, 229
Spectroscope, 12, 80
Spectroscopist, 84, 173, 200
Spectrum
analysis, 11, 12, 17
of sodium, 11
solar, 10, 11, 13, 14
stellar, 70
Spherical harmonic, 201, 202, 217, 219
Spinrad, Hyron, 166
Spitzer, Lyman, Jr., 145
Sputnik, 137
Standard candle, 255, 259, 263, 264, 266
St. Andrews, 155–157, 169
Star
binary, 6, 7, 17, 75, 105, 235, 238, 239, 244, 254
comparison, 196, 215
dwarf, 8, 14, 22, 28, 29, 55, 56, 59, 71, 85, 111, 112, 118, 123, 128, 144, 151, 156, 159, 161, 166, 186, 195, 221, 231, 237, 249, 256, 257, 259, 264, 268
formation, 61, 249, 250, 252, 257
giant, 235
hot subdwarf, 246
multiple, 237
pulsating, 199, 203, 206, 207, 259
target, 195, 214, 215
ZZ Ceti, 192, 198–200, 202–204, 207–209
Star cluster
globular, 149, 248
Hyades, 68
Praesepe, 254
Stark effect, 77
Starrfield, Sumner, 245
Stebbins, Joel, 69
Stefan-Boltzmann law, 12, 15, 22, 53, 171
Stefan, Josef, 11
Stein 2051 B, 172
Stellar
atmosphere, 77, 126, 128, 129, 151, 153, 154, 156, 165, 177, 181, 192, 229, 230
evolution calculations, 101–103, 106, 114, 117, 220, 237
nursery, 252
oscillation mode, 200, 202, 212
pulsation code, 198
winds, 161, 188
Stevenson, David J., 119
STIS. *See* Space Telescope Imaging Spectrometer (STIS)
Strand, Kaj Aa., 169

Stratified model, 200, 203
STRETCH, 97
Strittmatter, Peter A., 157, 184, 185
Strömgren, Bengt, 181
  UVBY colors, 181
Struve, Otto, 7
Subdwarf B variable (sdBV), 207
Sun
  central density, 19, 20
  central pressure, 19
  central temperature, 20, 44, 119
Supercomputer, 97
Supermassive black hole, 247
Supernova (SN)
  Type Ia (SN Ia), 246, 263–267
  Type Ib, 263
  Type Ic, 263
Supernova Cosmology Project, 264, 266
Supernovae (SNe), 245, 246, 263–267
Super outburst, 243
Surface gravity, 16, 128, 140, 150–152, 154, 160, 165, 173, 174, 181
SU Ursae Majoris, 243
Swan bands, 83, 84, 228

T

Tabulating Machine Company, 92
Taylor, Joseph, 244
Telescope
  100-inch, 79, 260
  200-inch (5-meter), 78–82, 140, 174
  Keck I, 266
  Keck II, 266
  Multiple-Mirror (MMT), 206
Television, 131, 137, 139
Teller, Edward, 111
Temperature, 7, 11, 17, 34, 42, 53, 70, 78, 87, 99, 109, 124, 140, 151, 165, 184, 195, 218, 229, 240, 252, 261
  gradient, 18, 25, 53, 55, 124, 125, 128, 133, 157, 177
Tensor, 31, 230
Terashita, Yoichi, 154
Tests of general relativity, 31
Texas Instruments, 72
Thejll, Peter, 161
Theoretical Division (at Los Alamos), 92
Thermal
  conduction, 55
  energy, 45, 55, 60, 111–113, 133, 265
  pulse (instability), 104, 106, 235, 242
  radiation, 11, 12, 24, 261, 262
Thermodynamic properties, 111, 131, 132
Thermodynamics, 27
Thermonuclear
  explosion, 102
  reaction, 44, 46, 104, 264, 265
  runaway, 104, 245, 265
Thermostat, 265
Thomson scattering, 163
Thomson, William, 17
Time series, 195, 211, 214–216
Tombaugh, Clyde W., 68
Transistor, 72, 96, 97
Trapped mode, 202–204
Trimble, Virginia, 80, 83, 174–176
Trinity College, 37, 59
Triple-alpha reaction, 51
Truran, James, 245
Tsiolkovsky, 138
Tutukov, Alexander, 246
Two-channel model, 183

U

U Geminorum, 242–244
Uhlenbeck, George, 33
Ultraviolet radiation (UV), 106, 139, 150
Uncertainty principle, 34
UNIVAC, 95, 98
Universe, 48, 137, 138, 146, 176, 183, 235, 250, 256, 259–268
University
  of Amsterdam, 63
  of Arizona, 77, 165, 177, 184, 206, 231, 249
  Australian National, 184, 232
  of California, Berkeley, 99, 165
  of California, Irvine, 82
  of California, Los Angeles (UCLA), 174
  Cambridge, 36, 59, 100, 157, 175, 184
  of Cape Town, 213
  of Chicago, 39, 75, 76, 78, 79, 82, 89, 95, 99, 117, 153
  College London, 102, 162
  of Colorado, 198, 226
  Columbia, 111
  of Copenhagen, 169
  Cornell, 43, 51, 54, 114
  de Chile, 251
  of Delaware, 156, 158, 171, 173, 179, 216, 250
  de Montröal, 130, 135, 160–163, 188, 250, 269
  George Washington, 44
  Harvard, 82
  of Illinois, 100, 101, 127, 239, 245

of Iowa, 137
Iowa State, 215
of Kansas, 165
Kiel, 126, 151, 157, 158
of Laval, 130
of Leicester, 147, 148
Leiden, 63, 75
Louisiana State, 159, 195
of Maryland, 83, 175
of Minnesota, 66, 69
of Mississippi, 5
of Oklahoma, 264
of Oregon, 225
Oxford, 37
of Paris, 123, 187
of Pennsylvania, 90, 91, 94, 96
Princeton, 21, 101, 114, 196
of Rochester, 57, 112, 115, 129, 130, 159–161, 198, 199, 204, 225
of Sussex, 61
of Texas, 173, 197–198, 201, 204, 212, 213, 215, 217
of Toronto, 101
Villanova, 85, 186
of Virginia, 196
Virginia Polytechnic and State, 57
of Warsaw, 101
of Washington (Seattle), 126, 127, 157–158, 177
of Western Ontario, 135
Yale, 4, 101, 153
Unsöld, Albrecht, 126, 151, 158
U.S. Army (Air Force), 89, 137, 169
U.S. Naval Observatory (USNO), 169, 170, 250, 252
U.S. Navy, 165, 212
UV. *See* Ultraviolet radiation (UV)

V
Vacuum-tube
amplifiers, 70
electronics, 96
technology, 72, 94
Van Allen, James A., 137
Vanguard, 137
van Maanen 2, 15, 64, 83, 85, 128, 151, 153, 177
van Maanen, Adriaan, 15
Vauclair, Gerard, 186–188, 200, 202, 209
Vauclair, Sylvie, 188
Vega, 12, 141
Vennes, Stefane, 161, 180
V477 Herculis, 205
Vidicon, 139
Vila, Samuel C., 114
Viscous energy dissipation, 242
Visual (wavelengths), 106, 144
Voltage, 72
von Braun, Wernher, 137
von Neumann, John, 93–96
Voyager (spacecraft), 141

W
Warner, Brian, 197, 198, 213, 244
Water, 113
Watson, Thomas J. Jr., 91, 94
Watson, Thomas J. Sr., 92
Wavelength calibration, 148
Weak interactions, 57
Weapons laboratory, 102
Webbink, Ronald F., 239, 245, 246
Weber, Joseph, 175
Wegner, Gary, 176–178
Weidemann, Volker, 128, 151–153, 158, 159, 176, 177, 181, 182, 200, 207, 249
Weigert, Alfred, 101, 238
Weizsäcker, Carl Friedrich von, 43
Wesemael, François, 83, 85, 139, 159–161, 207, 250
WET. *See* Whole Earth Telescope (WET)
White dwarf
age, 103, 111, 118, 119, 235, 249–251, 256
binary, 228
candidates, 75, 196
central density, 57, 111, 264
central temperature, 103, 119, 133–135
CO, 240, 243, 244, 246
cooling phases, 112
cooling track, 105, 106, 111, 170, 171, 192, 222
core temperature, 57, 112, 128, 135
density, 34
He, 84, 239, 243
instability strip, 199
isothermal core, 134
luminosity, 58, 119, 128, 166, 249, 251, 252
Luminosity Function (WDLF), 166, 249–252, 255–257
mass-radius relation, 171–173
ONe, 246
spectra, 75–86, 165, 177, 183, 184, 186, 191
spectral classification, 78
surface gravity, 128, 140, 152, 160, 173

White, Snow, 68
Whole Earth Telescope (WET), 173, 211–223
Wickramasinghe, Dayal T., 157, 184, 185, 228, 230–235
Wide Field Camera (WFC), 149
Wien displacement law, 12
Wien, Wilhelm, 12
Williams College, 100
Wilson, Robert W., 261
Winget, Donald E., 118, 200–202, 204, 217, 219, 221, 222, 257
Wood, Matt, 170, 172, 250, 251
Woolsthorpe, 9
World War I (WW I), 25, 78
World War II (WW II), 57, 68, 80, 82, 87, 89, 102, 123, 126, 168, 212, 261
Wu, Yanqin, 209

X

XCOV, 214, 216

Y

Yerkes, Charles T., 75, 76
York, Donald G., 139

Z

Zalitacz, Stefania, 114
Z Camelopardalis, 243
Zeeman splitting, 225–227, 231
Zwicky, Fritz, 68, 263
ZZ Ceti, 192, 197–200, 202–205, 207–209, 214
ZZ Ceti white dwarf, 197–199

Made in United States
Orlando, FL
08 June 2024